Laser-Solid Interactions
for Materials Processing

MATERIALS RESEARCH SOCIETY
SYMPOSIUM PROCEEDINGS VOLUME 617

Laser-Solid Interactions for Materials Processing

Symposium held April 25–27, 2000, San Francisco, California, U.S.A.

EDITORS:

D. Kumar

Center for Advanced Materials and Smart Structures
North Carolina A&T State University
Greensboro, North Carolina, U.S.A.

David P. Norton

University of Florida
Gainesville, Florida, U.S.A.

Clinton B. Lee

North Carolina A&T State University
Greensboro, North Carolina, U.S.A.

Kenji Ebihara

Kumamoto University
Kumamoto, Japan

X. X. Xi

The Pennsylvania State University
University Park, Pennsylvania, U.S.A.

Materials Research Society
Warrendale, Pennsylvania

CAMBRIDGE UNIVERSITY PRESS
Cambridge, New York, Melbourne, Madrid, Cape Town,
Singapore, São Paulo, Delhi, Mexico City

Cambridge University Press
32 Avenue of the Americas, New York NY 10013-2473, USA

Published in the United States of America by Cambridge University Press, New York

www.cambridge.org
Information on this title: www.cambridge.org/9781107413108

Materials Research Society
506 Keystone Drive, Warrendale, PA 15086
http://www.mrs.org

First published 2001
First paperback edition 2013

Single article reprints from this publication are available through
University Microfilms Inc., 300 North Zeeb Road, Ann Arbor, MI 48106

CODEN: MRSPDH

ISBN 978-1-107-41310-8 Paperback

Cambridge University Press has no responsibility for the persistence or
accuracy of URLs for external or third-party internet websites referred to in
this publication, and does not guarantee that any content on such websites is,
or will remain, accurate or appropriate.

CONTENTS

FUNDAMENTALS OF LASER
DESORPTION AND ABLATION

LASER-DRIVEN
NANOPARTICLE FORMATION

*Invited Paper

POSTER SESSION:
LASER-SOLID INTERACTIONS
FOR MATERIALS PROCESSING

*Invited Paper

LASER DIRECT WRITING

LASERS IN MICROMACHINING
AND SURFACE MODIFICATION

LASER-BASED DEPOSITION
OF OXIDES

*Invited Paper

PULSED-LASER DEPOSITION

PREFACE

This volume is the proceedings of the symposium "Laser-Solid Interactions for Materials Processing," held April 25-27 at the 2000 MRS Spring Meeting in San Francisco, California. The intent of the symposium was to bring together the materials science and technology communities to discuss recent progress on various aspects of laser ablation, which included both the fundamentals of pulsed laser deposition and its potential application in thin film formation and materials processing.

We would like to thank all of the participants and contributors from academia, national institutions and industrial organizations who made this symposium so successful. There were more than 75 papers presented in seven sessions focusing on (i) fundamentals of laser desorption and ablation, (ii) laser-driven nanoparticle formation, (iii) laser-solid interactions for materials processing, (iv) laser direct writing, (v) lasers in micromachining and surface modification, (vi) laser-based deposition of oxides, and (vii) pulsed-laser deposition. We would also like to acknowledge the financial contributions of Epion Corporation, Lambda Physik, and Neocera, Inc.

D. Kumar
David P. Norton
Clinton B. Lee
Kenji Ebihara
X. X. Xi

April 2000

MATERIALS RESEARCH SOCIETY SYMPOSIUM PROCEEDINGS

MATERIALS RESEARCH SOCIETY SYMPOSIUM PROCEEDINGS

Prior Materials Research Society Symposium Proceedings available by contacting Materials Research Society

Fundamentals of Laser
Desorption and Ablation

Mat. Res. Soc. Symp. Vol. 617 © 2000 Materials Research Society

When a Mild-Mannered 1-5 eV Photon Meets a Big 10 eV Bandgap:
Studies of Laser Desorption from Modified Surfaces of Ionic Single Crystals

J. Thomas Dickinson, Christos Bandis, and Stephen C. Langford
Surface Dynamics Laboratory, Washington State University, Pullman, WA 99164-2814

ABSTRACT

Exposing wide-bandgap ionic materials to UV and IR photons can produce ion emissions with kinetic energies of several eV, well in excess of the photon energy. Electron emissions are also observed. This implies that these materials possess occupied electronic defect states within the band gap. We have investigated the consequences of a variety of defect-generating stimuli (electron irradiation, laser irradiation, mechanical treatments, and heating) on electron and ion emission from inorganic ionic crystals. These stimuli generate defects that strongly interact with the probe laser on a wide variety of ionic crystals, and dramatically decrease the probe laser intensities required for ion and neutral emissions, laser damage, and plume formation.

INTRODUCTION

Laser ablation and desorption are often employed in conjunction with mass and optical spectroscopies for chemical analysis and materials characterization [1]. Surface modification, etching, and machining with lasers are also becoming increasingly important. Each of these operations requires detailed understanding of possible laser interactions, including those that alter subsequent laser interactions [2]. Electron and ion emission are of special interest due to their role in the onset of plasma formation,[3,4] damage, and large scale material removal. Considerable research has been devoted to the interaction of nanosecond laser pulses with sub-bandgap wavelengths with ionic insulators in order to determine the mechanisms by which electrons, positive ions, and neutral atoms and molecules are desorbed from the surface, and how these processes are affected by material history.

EXPERIMENT

Single crystals of $NaNO_3$ (99.0% pure) were grown by slow cooling from the melt and by evaporation from aqueous solution. Single crystals of a brushite (monoclinic $CaHPO_4 \cdot 2H_2O$), a model biomaterial, were grown by diffusion over time periods of several weeks from aqueous solutions of $Ca(NO_3)_2$ and $NH_4(H_2PO_4)$, adjusted to a pH of 2-4 with nitric acid [5,6]. Crystals were cleaved in air into ~2-mm thick slices and mounted into a vacuum chamber with base pressure of 1×10^{-9} Torr. Sample irradiation was performed with a Lambda Physik excimer laser (30 ns pulse width at 248 nm) and a Continuum SureLight II Nd:YAG laser (10 ns pulse width at 1064 nm). The intensity of the incident IR radiation (1064 nm) was varied with a CVI continuously variable beamsplitter, while the intensity of the UV radiation (248 nm) was varied with a Microlas Lasersystem variable attenuator. The emitted particles were detected with a UTI 100C quadrupole mass spectrometer (QMS) mounted with its axis along the sample surface normal. Mass resolution was better than ±1 amu/e. The mass filter was typically tuned to a specific mass/charge ratio and the output detected as a function of time. The output of the QMS detector was amplified by an Ortec Model 579 fast filter amplifier (rise time 5 ns). In the work on sodium nitrate, this signal was then digitized with a Lecroy LC584AXL, 1 GHz digital oscilloscope, and finally discriminated and counted with software (a peak finding routine). The resulting counts vs. time were summed over 50-200 laser pulses with a typical time resolution of 500 ns. In the work on brushite, the signals were discriminated and pulse counted in real time with a Princeton Applied Research Model 914P multichannel scaler. These two techniques yield entirely consistent time of flight (TOF) data. The use of software to identify individual ion counts in digitized signals (employed in the analysis of data from sodium nitrate) allows for a more direct computation of ion energy distributions that eliminates artifacts in the low energy region. Since software discrimination requires orders of magnitude more memory for data storage, hardware discrimination is generally more convenient.

RESULTS AND DISCUSSION

Sodium Nitrate

In previous work, we showed that single crystal $NaNO_3$ yields strong ion emissions (Na^+, Na_2^+, Na_2O^+) when irradiated at 248 nm. Significant emissions of neutral NO and O_2 are also observed, which we attribute to dissociative absorption into the long wavelength tail of the $\pi \rightarrow \pi^*$ transition within the nitrate anion, centered at 6.4 eV [7]. The kinetic energies of the resulting ions were centered at about 7 eV, well above the energy of the incident photons. We proposed that the emitted ions were originally adsorbed atop surface electron traps, and that the sequential absorption of several UV photons ionized the underlying traps. Simple point charge models of the $NaNO_3$ lattice indicate that the net force on the adsorbed ions becomes strongly repulsive when the underlying electron trap is photoionized. This repulsion is sufficient to account for the observed kinetic energies [8,9]. The sequential photon absorption required for emission also accounts for the highly nonlinear fluence dependence of the emission intensities.

Given the large bandgap of $NaNO_3$ (> 6 eV), the photoionization events envisioned in the proposed emission model require localized defect states within the band gap. To explore more precisely where these states might be, we explored the ion emissions accompanying irradiation at lower photon energies, including the 1.16 eV photons at 1064 nm. Single crystal $NaNO_3$ is much more transparent at 1064 nm than at 248 nm. Nevertheless, 1064-nm radiation at fluences well below the single pulse damage threshold (~4 J/cm^2) produces both electron and Na^+ emissions. These emissions typically disappear over the course of prolonged exposure at 1064 nm. However, they are restored by brief exposures to UV radiation (248 nm) or energetic electrons (2.5 keV), both of which are know to create both lattice and electronic defects in this material [10,11]. Sodium nitrate is of interest as a model environmental material and a component of many nuclear wastes.

When freshly cleaved $NaNO_3$ is first exposed to IR radiation, the resulting positive ion signal is both intense and energetic. A typical ion time-of-flight (TOF) signal acquired during the first 200 laser pulses at a fluence of 0.25 J/cm^2 on a previously unexposed surface is shown in Fig. 1. Quadrupole mass selected measurements at other mass settings yield no significant signals. In contrast to ion emissions under 248-nm radiation, 1064-nm radiation yields only Na^+ at this level of sensitivity. The kinetic energies corresponding to several arrival times are indicated at the top of the diagram. The large number of ions with energies greater than 1 eV is inconsistent with thermal emission mechanisms.

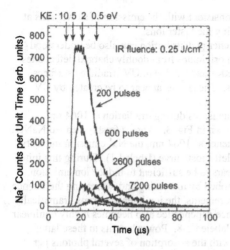

Figure 1. Ion TOF distributions acquired during 200 pulses at 1064 nm at a fluence of 0.25 J/cm², where the total number of accumulated pulses at the end of each data acquisition are indicated for each curve.

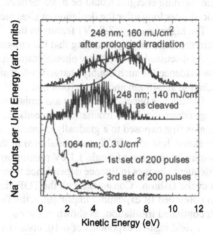

Figure 2. Na⁺ ion energy distributions accompanying IR irradiation of NaNO₃ due to IR (1064 nm) and UV (248 nm) laser radiation on freshly cleaved surfaces, and after prolonged laser exposure.

Also shown are TOF signals acquired during subsequent exposures, each 200 pulses long at the same fluence. As irradiation proceeds, the emission grows much weaker and slows down. The gradual weakening of the signal indicates the depletion of sites responsible for emission. Assuming that each site participating in emission yields ions with a distinct kinetic energy, the shift to lower kinetic energies suggests that at least two kinds of sites are responsible for emission, and that the higher kinetic energy sites are depleted more rapidly.

The effect of depletion on the ion TOF is readily apparent in the ion kinetic energy distributions, several of which are plotted in Fig. 2. These energy distributions were formed by recording the time of arrival for thousands of individual ions and counting the number of ions in each energy bin. This process avoids potential low energy artifacts that can be introduced when transforming TOF data. Also plotted are energy distributions determined in a similar fashion during UV (248 nm) laser irradiation. The initial energy distributions for ions from freshly cleaved surfaces under both UV and IR show a distinctive feature at 4 eV. In the case of IR radiation, this energy is much greater than the incident photon energy.

The energy of the 4 eV feature can be accounted for by assuming that the emitted ions are electrostatically ejected from surface sites on top of ionized electron traps, where the nearby charges are treated as point charges. The energy of the departing ion is then approximately equal to the electrostatic repulsive energy minus the Madelung binding energy, BE, of ions adsorbed at specific surface sites. Assuming that all ions begin at ideal lattice positions one nearest neighbor distance above a photoionized electron trap, and neglecting relaxation effects, the resulting ion kinetic energy, KE, is 4.1 eV ($KE = e^2/r - BE$; $r = 3.24$ Å, $BE = 0.3$ eV). Within the context of this model, slower ions would originate from sites situated at some distance from the ionized defect, or from sites with higher binding energies (i.e. cleavage steps). Emission from sites with higher binding energies would be more difficult, consistent with the emission of slower ions at reduced rates long after the depletion of sites yielding the faster ions.

The higher energy (7-8 eV) feature in the UV energy distribution can also be understood within the same model, assuming that the emission originates from doubly charged defect sites [8]. As discussed below, X-ray photoelectron spectroscopy (XPS) of UV-irradiated surfaces show evidence for such doubly charged defect sites. These sites appear to be created by UV irradiation at 248 nm and not at 1064 nm.

The fluence dependence of the ion emission intensities during irradiation at 1064 nm changes dramatically as irradiation proceeds, as shown in Fig. 3. When a freshly cleaved $NaNO_3$ crystal is first exposed to a gradually increasing fluence at 1064 nm, the Na^+ emission intensity rises almost linearly with fluence, as shown in the left-most curve (labeled 1). During this initial exposure, absorption of single 1.1 eV photons appears to be sufficient to induce ion emission. Subsequent changes in fluence dependence were probed by increasing and decreasing the fluence in a cyclic fashion. Consistent with the TOF measurements, the ion intensity drops dramatically with continued IR exposure. Importantly, the fluence dependence also becomes highly nonlinear with continued irradiation, as shown in the curves labeled 2-8. Power law fits to these latter curves yield best-fit exponents of 6-10, consistent with the absorption of several photons per emission event. Thus the depletion of sites yielding initial emission not only lowers the kinetic energy, but also raises the order of the fluence dependence. Although these observations are consistent with the previously proposed emission mechanism, the initial linear fluence dependence is surprising and implies the presence of shallow electron traps.

Thus the freshly cleaved $NaNO_3$ surface appears to have weakly bound ions that yield energetic (up to at least 4 eV) ions with the absorption of as few as one 1.1 eV photon. The simplest scenario for this emission involves metastable bound ions. Simple point charge models of the emission process indicate that a positive ion adsorbed on top of an electron trap will remain bound as long as the underlying trap retains its negative charge, even after the ionization of one or more nearby (even next-nearest neighbor) electron traps. This allows the net binding energy of the ion to be quite negative without emission and permits sequential photoionization events involving several individual photons to contribute to the negative binding energy before emission occurs [12]. When only one photon is required for emission of an energetic ion (kinetic

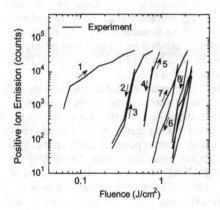

Figure. 3. *Fluence dependence of Na^+ ion yield from as-cleaved $NaNO_3$ accompanying 1064 nm laser irradiation, where the fluence was successively increased and decreased. The order in which the data were taken is indicated by the numbers (1 = initial fluence dependence measurement on a freshly cleaved surface) and the arrows. After initial exposure, the fluence dependence becomes highly nonlinear and the ion yield decreases rapidly.*

energy greater than the photon energy), the adsorbed ion must be initially in a metastable state. Cleavage often produces localized patches of surface with significant net charge densities [13,14] and rearranges the surface material in ways that could easily populate the surface with adsorbed ions [15]. Metastable adsorbed ions in such patches would be ideally situated for subsequent emission under 1064-nm radiation, yielding energetic emissions with a weak fluence dependence. The number of such ions, however, would be limited. Subsequent emission would involve more tightly bound ions, requiring additional absorption events to induce emission, and yielding lower kinetic energy emissions.

Electron emission (not shown here) is also observed upon irradiation of as cleaved $NaNO_3$ at 1064 nm [16]. Like the ion emission, the fluence dependence of the electrons is highly nonlinear after exposure to several hundred laser pulses at 1064 nm. We suggest that these electron emissions involve the sequential absorption of several IR photons during each laser pulse, similar to the ion emissions.

The depletion accomplished by prolonged IR exposure can be largely reversed by exposure to intense UV (248 nm). Previous studies found that 248-nm irradiation produces a defect-rich, sodium-rich surface that yields intense ion and neutral emissions [8,17,18]. Thus UV radiation has the potential to replace defects consumed by ion emission during IR exposure. Figure 4 shows the fluence dependence of ion emission accompanying irradiation at 1064 nm after prolonged IR exposure (curves 1, 2) and subsequent ion emission accompanying 1064-nm radiation when the surface was exposed to one 248-nm laser pulse between each IR laser pulse (curves 3, 4). We emphasize that the Na^+ intensities plotted here are generated by the IR laser, not by the UV laser. Curves 3 and 4 show that exposing the surface to 248-nm radiation between IR pulses dramatically increases the Na^+ intensities accompanying the IR pulses. In terms of our ion emission model, exposure to UV has increased the density of the electron-trap/adion defect complexes that yield emission during IR exposure.

XPS measurements of UV-irradiated surfaces confirm that exposure produces high densities of surface and near-surface defects. Typical spectra of UV-irradiated surfaces are shown in Fig. 5, along with spectra of freshly cleaved $NaNO_3$ and $NaNO_2$ for comparison. Surfaces exposed to 248-nm radiation (the bottom curve in each diagram) show pronounced features in the N 1s spectrum that can be attributed to nitrite ions (the middle curve in each spectrum). The

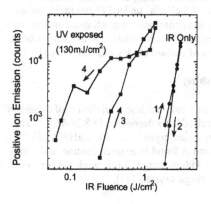

Figure. 4. *Fluence dependence of the Na+ ion emission from NaNO₃ accompanying IR (1064 nm) laser pulses; between each pair of IR pulses, the surface was exposed to one UV (248nm) laser pulse. The arrows indicate low the IR fluence was varied and the numbers indicate the order the data were collected. Exposure to UV laser radiation significantly increases the IR-induced ion intensities.*

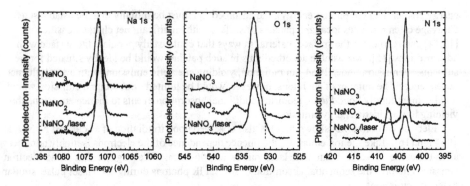

Figure 5. *XPS core level spectra of as-cleaved, melt grown sodium nitrate (NaNO₃), as cleaved, melt grown sodium nitrite (NaNO₂), and cleaved NaNO₃ after UV excimer laser irradiation (NaNO₃/laser). Exposure of the as cleaved NaNO₃ surfaces to both x-rays and UV laser radiation produces principally sodium nitrite and sodium peroxide.*

production of nitrite species in nitrates by X-ray and UV irradiation has been well studied [19,20]. In addition, the O 1s spectrum shows a low binding energy feature at 531 eV that can be attributed to O_2^{2-} ions [21]. The presence of doubly charge species on the irradiated surface is consistent with the presence of doubly charged surface electron traps, and helps account for the emission of 7-8 eV Na^+ ions after exposure to 248-nm radiation.

These results are also consistent with recent studies [8,17,22,23] of the neutral emissions during UV laser radiation. Upon irradiation, $\pi \rightarrow \pi^*$ transitions result in decomposition of the nitrate NO_3^{*-} to either $NO(g) + O_2^-(s)$ or $NO_2^{*-} + O(g)$, with the possibility of further decomposition to $NO(g) + O^-$ (s), where (g) and (s) denote gas and surface species, respectively. Surface oxygen can be emitted as O_2 or react to form oxide and peroxide species (doubly charged defects) on the surface. Neutral NO and O_2 emissions are observed during 248-nm irradiation [17]. In addition, the ionic species Na_2O^+ and $Na_2O_2^+$ have been observed [17]. Both oxides and peroxides are doubly charged defects in the near surface region. Further, the emission of NO and O_2 may leave anion vacancy defects at singly and doubly charged defect sites. Some of these sites could yield desorption of 6-8 eV Na^+ under 248-nm radiation, as discussed above. In agreement with this scenario, ion emissions are much more intense from surfaces which have been previously exposed to either 248-nm radiation or to electron beams, and which, therefore, exhibit higher concentrations of sodium oxides and peroxides [10,18].

Calcium Hydrogen Phosphate Dihydrate (Brushite)

The interaction of minerals and biomaterials with excimer laser radiation is of considerable for chemical analysis, surface modification, and pulsed laser deposition [3,8,24-26]. Many minerals and biomaterials contain water, including single crystal brushite ($CaHPO_4 \cdot 2H_2O$). Brushite is a precursor material for bone growth, and is found in enamel, dentine, and some pathological structures, such as dental cavities and kidney stones. Little work has been done on laser interactions with water-containing inorganic single crystals [27].

Brushite is a layered monoclinic crystal with alternating sheets of CaHPO$_4$ and water molecules parallel to the (010) cleavage surface [28-30]. As grown, these crystals are highly transparent. Rapid dehydration and internal recrystallization during heating produces voids which render them opaque throughout their bulk [31]. In this work, we show that these thermally induced changes have a profound affect on UV-induced ion desorption from brushite surfaces, which we attribute to the production of defects which interact strongly with UV laser radiation [12].

Ion TOF distributions obtained from brushite are consistent over a broad range of fluences, ruling out fluence dependent effects such as surface charging as a source of kinetic energy. Figure 6 shows TOF distributions of Ca$^+$ desorbed from a rough region of a heated-cleaved surface at three fluences from 0.7 to 1.1 J/cm^2. We attribute the double-peaked structure of TOF distributions from brushite to emission from two distinct surface sites. As in NaNO$_3$ under UV radiation, the ion intensity grows rapidly with fluence.

The ion intensities are increased by a variety of treatments, including heating. In general, these treatments do not affect the TOF distributions significantly. Figure 7 shows UV induced Ca$^+$ TOF signals from: (a) an as-cleaved surface, (b) a smooth region on a heated-cleaved

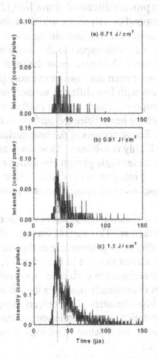

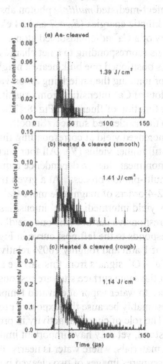

Figure 6. Ca$^+$ mass selected TOF signals from UV irradiated, heated-cleaved CaHPO$_4$·2H$_2$O at fluences of (a) 0.71 J/cm^2, (b) 0.91 J/cm^2, and (c) 1.1 J/cm^2..

Figure 7. Ca$^+$ mass selected TOF signals from UV irradiated CaHPO$_4$·2H$_2$O surfaces: (a) as-cleaved, (b) a heated-cleaved smooth region, and (c) a heated-cleaved rough region. Solid lines show least-squares fits of Eq. (1) to the data.

surface, and (c) a rough region on a heated-cleaved surface. Three different fluences were used to produce TOF distributions with approximately the same peak intensities. As shown by the dotted lines, the three signals each have peaks at about 34 µs and 48 µs, independent of surface treatment. These signals are similar to reported Ca^+ signals from as-grown brushite [27].

These TOF signals are readily described in terms of TOF distributions corresponding to the sum of two Gaussian energy distributions:

$$I(t) = \sum_{i=1}^{2} \frac{A_i m d^2}{t^3} \cdot \exp[\frac{-(E_i - \frac{md^2}{2t^2})^2}{2\sigma_i^2}] \qquad (1)$$

where d is the distance between the sample surface and the CEM (28 cm), m is the mass of the ion, t is the ion time-of-flight flight, E_i is the mean kinetic energy, σ_i is the standard deviation of the kinetic energy, A_i is a constant and $i = 1,2$. The solid lines in Figs. 6 and 7 show the result of curve fits to Eq. (1), with best fit kinetic energies, E_i, corresponding to roughly 7 and 14 eV.

Both kinetic energies exceed the 5 eV UV laser photon energy, and can be explained in terms of the defect-mediated *multiple* photon absorption process discussed above [8,9,12]. The most probable kinetic energies (E_i, ≈ 7 and 14 eV) correspond closely to the initial electrostatic potential energy of a Ca^+ adion adsorbed atop a surface anion vacancy containing one or zero trapped electrons (corresponding to a net charge of $+e$ or $+2e$ with respect to the ideal lattice), and the adion is positioned one bulk nearest neighbor distance (0.33 nm) above the vacancy [32]. Mechanisms for ionizing the underlying anion vacancy have been discussed previously [8,9,33].

Log-log plots of Ca^+ intensities from brushite surfaces with four different surface treatments are shown as a function of fluence in Fig. 8. The least responsive surface, data marked by (▲), corresponds to the as-cleaved surface. Smooth regions of the heated/cleaved surface, marked by (●), show an increased yield at corresponding fluences. Rough regions of the heated/cleaved surface show still higher yields (▼). All four curves are highly non-linear. As we have argued earlier,[8] the nonlinear fluence dependence requires *multiple* single-photon absorption events, rather than a multi-photon process. At a given fluence, the rough regions of heated/cleaved surfaces yield 2-4 orders of magnitude more Ca^+ than as-cleaved surfaces; smooth regions of heated surfaces yield intermediate Ca^+ intensities.

The most responsive surface (the one yielding the most intense emissions at the lowest fluences) corresponds to the leftmost curve in Fig. 7, marked ■. This surface was heated at 110 °C for 5 hours at 1 atm and roughly 100% relative humidity prior to mounting in the vacuum system. The Ca^+ TOF signals from this surface are very similar to those in Figs. 6 and 7, suggesting that the same surface sites are responsible for emission in each case. Exposing brushite surfaces to water vapor at elevated temperatures dramatically increases the ion emission intensities, presumably because this exposure produces high densities of the defects responsible for emission. The role of chemical activity in producing defect sites for emission processes is not well understood, yet may have important implications in the ablation of biological and environmental materials, where water is nearly ubiquitous.

Figure 9 compares images of two cleaved brushite surfaces, one heated in vacuum (rough region) and the other heated at 110 °C for 5 hours at 1 atm and roughly 100% relative humidity. Both surfaces yielded intense ion emission. The rough, recrystallized material in the center of Fig. 9(a) is virtually indistinguishable from rough regions on brushite surfaces formed by

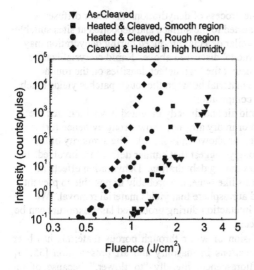

Legend:
▼ As-Cleaved
■ Heated & Cleaved, Smooth region
● Heated & Cleaved, Rough region
◆ Cleaved & Heated in high humidity

Figure 8. Fluence dependence of Ca⁺ intensities from brushite for four different surfaces treatments: (▲) as cleaved; (●) heated-cleaved smooth region; (▼) heated-cleaved rough region; and (■) cleaved and heated at high humidities.

cleaving through samples heated in dry air. In both cases, heating induces dehydration and recrystallization *beneath* the surface. The surface material itself is dehydrated by exposure to dry air or vacuum and forms a thin, relatively impermeable layer. The recrystallized material below is exposed only when this impermeable layer is broken, for instance, by pressurized vapor accumulating beneath the surface layer or by deliberate cleavage during sample preparation. In cases where rough regions become exposed during the course of laser irradiation, the ion emissions increase dramatically [34].

Although the composition of the rough, recrystallized material is not precisely determined, IR absorption of fragments removed from heated surfaces is consistent with simple dehydration. Apart from any expected enhancements due to surface roughening, the intense emission from

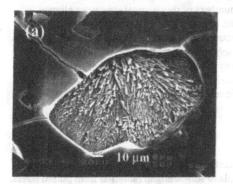
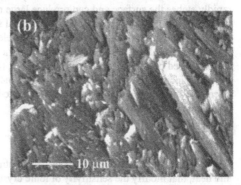

Figure 9. SEM images of (a) an exposed cavity with recrystallized platelets on a brushite surface heated in a vacuum, and (b) an as-cleaved brushite surface heated to 110 °C at about 100% relative humidity.

these rough patches appears to be related to the process of dehydration. As water diffuses through cracks in the bulk, Ca^+ ions are transported to the surface and deposited at sites suitable for emission. Surface anion vacancies (and similar defects) produced during dehydration may play the role of electron traps in the emission model discussed above. Rapid recrystallization often produces highly defective material, consistent the high defect densities on the rough patches. In contrast, the less defective surface material between the rough patches yields much weaker emissions, despite a similar chemical composition.

Although brushite surfaces heated in humid air and directly mounted in vacuum lack the angular crystallites of those heated in vacuum or in dry air, they show strong evidence for complex recrystallization on micron size-scales, as shown in Fig. 9(b). The similarity of the emissions from these two types of surface strongly suggest that similar defects are involved, and that these defects are produced by ion transport during dehydration. Dehydration effects in hydrated minerals, including biological minerals like bone, are probably impossible to prevent during laser irradiation at any wavelength and atmosphere that yields material removal. We suggest that the strength of the laser-material interaction during prolonged laser exposure can be strongly affected by dehydration-induced defects.

Ionic transport associated with the diffusion of water through porous materials has been studied extensively and has important implications in masonry and art preservation [35,36]. Initial investigators termed this process efflorescence, literally "to flower," because of the flower-like structures often produced on surfaces from which the water evaporates. The process of dehydration discussed above is quite similar, and it is perhaps not surprising that similar structures can be produced when the laser intensity is adjusted to fracture the surface of a hydrated material, without melting. Figure 9 shows an SEM image of a particularly beautiful efflorescence on a brushite surface which grew *after* 100 pulses of 248 nm radiation at 1.5 J/cm^2. This illustrates the power of diffusion to transport significant quantities of material.

High energy particles, of course, can have even more dramatic effects. When single crystal brushite is exposed to 2 keV electrons, electron-induced heating produces patches of recrystallized material that are virtually identical to those produced by heating without electrons [Fig. 9(a)]. Similar enhancements in ion emission are observed during laser irradiation. However, if the surface is heated sufficiently (e.g., by a resistance heater), electron irradiation rapidly etches the surface and can produce large numbers of thin conical structures aligned parallel to the incident electron beam. A typical set of cones formed during electron irradiation at about 530 °C is shown in Fig. 10, where the electron beam was incident at about 30° to the surface normal. Most cones show evidence for a spherical cap at the tip. Assuming these caps are composed of a refractory material like CaO, they would protect the underlying cone bodies from the electron beam while the surrounding material is etched. These conical structures are of interest in their own right and are under separate investigation as an example of electron stimulated desorption.

CONCLUSION

We have presented a variety of surface treatments, such as exposure to energetic photons and heat, that modify the sensitivity of ionic crystals to sub-bandgap radiation from pulsed lasers. These treatments dramatically decrease the laser intensities required for the ejection of ions and neutrals, promote plume formation, and render the materials more vulnerable to subsequent laser

 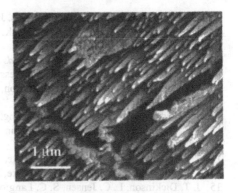

Figure 9. SEM image of efflorescence on as-grown brushite exposed to 100 pulses of 248-nm radiation at 1.5 J/cm².

Figure 10. SEM image of a brushite surface heated to approximately 530 °C and then exposed to a 0.05 C/cm² dose of 2 keV electrons.

radiation. Many of these effects are observed in other less complex materials such as MgO and NaCl. The creation of defects by one stimulus can create absorption centers appropriate for a second stimulus (laser radiation), as well as produce sites for the launching of adions. Reproducible surface modification, chemical analysis, and laser ablation will require the control of these effects. Similarly, these effects may be exploited to improve the localization of laser etching processes and to provide quantitative tools for the further study of laser-materials interactions.

ACKNOWLEDGMENTS

We thank Yoshizaki Kawaguchi, Chugoku National Industrial Research Laboratory, Japan, and Mary Dawes, Washington State University, David Ermer, Vanderbilt University, J.-J. Shin, National Taiwan University, Wayne Hess and Tom Orlando of Pacific Northwest Laboratory for their assistance and helpful discussions. This work was supported by the Department of Energy under Contract DE-FG03-99ER14864 and an Equipment Grant from the National Science Foundation, DMR-9503304.

REFERENCES

1. L. J. Radziemski and D. A. Cremers, *Laser Induced Plasmas and Applications* (Marcel Dekker, New York, 1989).
2. *Laser Ablation: Principles and Applications*; Vol., edited by J. C. Miller (Springer-Verlag, Berlin, 1994).
3. D. R. Ermer, S. C. Langford, and J. T. Dickinson, J. Appl. Phys. **81**, 1495-1504 (1997).
4. J. J. Shin, D. R. Ermer, S. C. Langford, and J. T. Dickinson, Appl. Phys. A **64**, 7-17 (1997).
5. R. Z. LeGeros and J. P. LeGeros, J. Crystal Growth **13/14**, 476-480 (1972).
6. R. Z. LeGeros, D. Lee, G. Quirolgico, W. P. Shirra, and L. Reich, in *Scanning Electron Microscopy*; Vol. *1993* (SEM Inc., Chicago, IL, 1983), p. 407-418.

7. H. Yamashita and R. Kato, J. Phys. Soc. Jpn. **29,** 1557-1561 (1970).
8. D. R. Ermer, J.-J. Shin, S. C. Langford, K. W. Hipps, and J. T. Dickinson, J. Appl. Phys. **80,** 6452-6466 (1996).
9. J. T. Dickinson, in *Experimental Methods in Physical Sciences*; *Vol. 30*, edited by J. C. Miller and R. F. Haglund (Academic Press, 1998), p. 139-172.
10. J.-J. Shin, S. C. Langford, J. T. Dickinson, and Y. Wu, Nucl. Instrum. Meth. Phys. Res. B **103,** 284-296 (1995).
11. J. T. Dickinson, J.-J. Shin, and S. C. Langford, Appl. Surf. Sci. **96-98,** 326-331 (1996).
12. J. T. Dickinson, S. C. Langford, J. J. Shin, and D. L. Doering, Phys. Rev. Lett. **73,** 2630-2633 (1994).
13. J. Wollbrandt, U. Brückner, and E. Linke, Phys. Status Solidi (a) **77,** 545-552 (1983).
14. J. Wollbrandt, U. Brückner, and E. Linke, Phys. Status Solidi (a) **78,** 163-168 (1983).
15. J. T. Dickinson, L. C. Jensen, S. C. Langford, and J. P. Hirth, J. Mater. Res. **6,** 112-125 (1991).
16. C. Bandis, S. C. Langford, J. T. Dickinson, D. R. Ermer, and N. Itoh, J. Appl. Phys **87,** 1522 (2000).
17. R. L. Webb, S. C. Langford, and J. T. Dickinson, Nucl. Instrum. Meth. Phys. Res. B **103,** 297-308 (1995).
18. C. Bandis, L. Scudiero, S. C. Langford, and J. T. Dickinson, Surf. Sci. **442,** 413-419 (1999).
19. L. K. Narayanswamy, Trans. Faraday Soc. **31,** 1411-1412 (1935).
20. E. R. Johnson, *The Radiation-Induced Decomposition of the Inorganic Molecular Ions* (Gordon and Breach, New York, 1970).
21. S. Aduru, S. Contarini, and J. W. Rabalais, J. Phys. Chem. **90,** 1683-1691 (1986).
22. W. P. Hess, K. A. H. German, R. A. Bradley, and M. I. McCarthy, Appl. Surf. Sci. **96-98,** 321-325 (1996).
23. K. Knutsen and T. M. Orlando, Phys. Rev. B **55,** 13246-13252 (1997).
24. R. L. Webb, L. C. Jensen, S. C. Langford, and J. T. Dickinson, J. Appl. Phys. **74,** 2323-2337 (1993).
25. R. L. Webb, L. C. Jensen, S. C. Langford, and J. T. Dickinson, J. Appl. Phys. **74,** 2338-2346 (1993).
26. J. T. Dickinson, L. C. Jensen, R. L. Webb, M. L. Dawes, and S. C. Langford, J. Appl. Phys. **74,** 3758-3767 (1993).
27. M. Dawes, S. C. Langford, and J. T. Dickinson, Appl. Surf. Sci. **127-129,** 81-87 (1998).
28. N. A. Curry and S. W. Jones, Chem. Soc. (London) A **1971,** 3725-3729 (1971).
29. M. Ohta, M. Tsutsumi, and S. Ueno, J. Crystal Growth **47,** 135-136 (1979).
30. M. Ohta and M. Tsutsumi, J. Crystal Growth **56,** 652-658 (1982).
31. H. Tanaka, N. Koga, and A. K. Galwey, J. Chem. Ed. **72,** 251-256 (1995).
32. Y. Kawaguchi, M. L. Dawes, S. C. Langford, and J. T. Dickinson, J. Appl. Phys. **88** (2000) (in press).
33. C. Bandis, S. C. Langford, and J. T. Dickinson, Appl. Phys. Lett. **76,** 421 (2000).
34. Y. Kawaguchi, M. L. Dawes, S. C. Langford, and J. T. Dickinson, Appl. Phys. A **69,** S621-S624 (1999).
35. A. G. Verduch, V. Sanz, J. V. Agramunt, and V. Beltrán, Am. Ceram. Soc. Bull. **75,** 60-64 (1996).
36. E. Ordonez and J. Twilley, in *Anal. Chem.*; *Vol. 69* (1997), p. 416A-422A.

Mat. Res. Soc. Symp. Vol. 617 © 2000 Materials Research Society

LASER ABLATION OF SOLID OZONE

Hidehiko Nonaka, Tetsuya Nishiguchi,[1] Yoshiki Morikawa,[1] Masaharu Miyamoto,[1] and Shingo Ichimura[2]

Materials Science Division and [2]Frontier Technology Division
Electrotechnical Laboratory, 1-1-4 Umezono,
Tsukuba, Ibaraki, 305-8568, Japan.
[1]Meidensha Corp. 515 Kaminakamizo Higashimakado,
Numazu, Shizuoka 410-8588, Japan.

ABSTRACT

Species ablated from solid ozone by a UV laser were investigated using a time-of-flight method through a quadrupole mass filter. The results show that UV-laser ablation of solid ozone can produce a pulsed ozone beam with a translational energy far above that of room temperature. High-concentration ozone from an ozone jet generator is solidified on a sapphire substrate attached to a copper block which is cooled to 30 to 60 K on a cryocooler head and the solid ozone is irradiated by pulsed laser light from a KrF laser (248 nm). The ablated species were a mixture of ozone and molecular oxygen as well as atomic oxygen due to photodissociation of ozone. At a substrate temperature of 30 K, the total amount of ablated ozone increases as the laser fluence increases to 13 mJcm^{-2}. Beyond this fluence, enhanced decomposition of ozone occurs. Gaussian fitting of the time-of-flight signals of the ablated ozone reveals an average thermal energy exceeding 1,500 K. The velocity also increases when the laser fluence enters saturation at 2,300 K at 13 mJcm^{-2}.

INTRODUCTION

Ozone, O_3, has the second largest oxidation potential (+2.076 eV) next to fluorine (+2.866 eV). [1] It has been predominant as an oxidizing agent, for instance, in preparing oxide superconductor films by reactive evaporation of metals in an ozonic atmosphere of less than 1×10^{-3} Pa. [2] In contrast, the pressure has to be at least 10^{-2} Pa in an oxygen atmosphere. [3] Furthermore, ozone is in the gas phase throughout the practical range of temperature and leaves only oxygen gas as a by-product of the oxidation reaction.

Because of ozone's advantageous properties, we have been investigating the potential of ozone, especially for application to the semiconductor device process in which extremely high integration of the devices requires an ultrathin, high-quality gate oxide that may no longer be fabricable using conventional processes in atmospheric oxygen. Using the high-purity ozone jet generator we developed [4], we have so far shown that silicon dioxide (SiO_2) films can be grown with low-pressure ozone (8×10^{-4} Pa) at a much lower temperature than with oxygen. [5] Furthermore, analysis of the oxide-silicon interface using medium-energy ion-scattering spectroscopy revealed that ozone-formed oxide is more homogenous than conventional oxide because silicon atoms near the interface have much smaller displacement in the ozone-formed oxide [6]. Thus, ozone is no doubt a promising agent in the device process. We are therefore trying to increase the advantages of ozone by producing a collimated hyperthermal ozone beam by laser ablation of solidified ozone. Laser ablation of solidified gas molecules is reported to produce hyperthermal gas beams for NO_2 [7] and Cl_2 [8]. Ozone molecules with higher kinetic energy may work as a stronger oxidizing agent against silicon to produce SiO_2 films at even a lower temperature. Furthermore, the collimated ozone beam could be used for new device processes such as local oxidation of the bottom part of a trench.

In this study, we employ laser ablation of solid ozone to produce an ozone beam. The velocity measurement of the ablated species shows that ozone as well as oxygen atoms and molecules are ejected as hyperthermal beams.

EXPERIMEMNTAL DEATAILS

A schematic diagram of the experimental setup is shown in figure 1. High-purity ozone gas (99 vol.%) from the ozone jet generator was introduced through an electrochemically polished stainless steel pipe onto a 1-inch diameter U V-grade sapphire substrate attached to a cryocooler head via a copper block with an indium sheet. The substrate could be cooled to 30 K, well below the melting point of ozone, i.e., 80 K [9]. The substrate was located in an ultrahigh vacuum chamber with a base pressure of 10^{-6} Pa. The chamber was equipped with a quadrupole mass filter (QMF) facing the substrate. Though the factors that may have lowered the purity of ozone were eliminated as much as possible, the purity of ozone gas at the substrate was estimated to be <80 vol.% because thermal

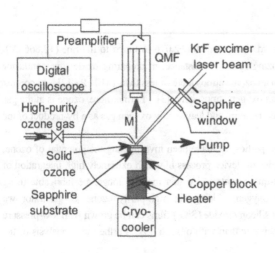

Figure 1 Schematic diagram of the experimental setup for laser ablation of solid ozone and analysis of ablated species M.

The distance between the sapphire substrate and the QMF ionizer was 25 cm.

decomposition of ozone into oxygen on the inner wall of the gas line pipe at room was unavoidable. Before laser irradiation, ozone was solidified on the substrate in a 100 μm thick film. This was estimated from the amount of ozone gas supplied to the substrate and is much larger than the estimated absorption depth into solid ozone at 248 nm, i.e., 1 μm [10]. The purity of the solidified ozone was estimated from the QMF signal intensity ratio of ozone to oxygen molecules when the substrate temperature, T_s, was slowly raised to vaporize the solid on the substrate. The purity was higher when ozone was solidified at a higher T_s, reaching over 80% at $T_s = 60$ K. For simplicity, the term "solid ozone" is used hereafter instead of "solidified ozone," though it is in fact a mixture of ozone and oxygen unless T_s is as high as the melting point of ozone.

The solid ozone on the substrate was irradiated through a UV-grade sapphire window with a pulsed laser light from a KrF excimer laser at an incident angle of 45°. The laser fluence at the substrate position was varied between 5 and 25 mJcm^{-2} using a quartz lens and an adjustable slit to maintain a constant irradiation area of 4 mm x 4 mm. The ablated species were mass-selected using the QMF, and their time of flight (TOF) between the substrate and the QMF ionizer was measured by putting the collector current of the QMF through a fast preamplifier into a digital oscilloscope triggered by the laser shot. The TOF through the QMF to the collector is negligible because of an intense acceleration of ions by the high-voltage static field. The signal was accumulated for 100 laser shots on a constant irradiation position on the solid ozone.

RESULTS AND DISCUSSION

The KrF laser light (248 nm) hits the Hartley bands of ozone at its maximum absorption coefficient region [10] resulting in photodissociation of ozone into an oxygen molecule and an oxygen atom in excited states with a quantum yield of about 2.0 [11] as shown in equation 1 [12].

$$O_3 + h\nu \ (248 \ nm) \ \rightarrow \ O_2 \ (a^1\Delta_g) + O(^1D) + 98.3 \ kJmol^{-1}. \qquad (1)$$

Therefore, the ablated species, if any, are expected to be neutral O_3, O_2, and O, unless the laser fluence is large enough to cause multiphoton ionization (MPI) of these species. No signal above the background was detected without emission of the filament of the QMF ionizer at the laser fluence of 13 mJcm^{-2}, and thus MPI would not take place within this fluence.

Figure 2 shows a typical QMF-TOF spectrum for m/e = 48 (O_3) at a T_s of 60 K and laser fluence of 13 mJcm^{-2}. Ozone molecules at the onset are calculated to have a translational energy exceeding 2 eV. A calculated TOF spectrum using equation 2 based on the Maxwell-Boltzmann (M-B) distribution at 1,500 K is also plotted.

$$f(t) \ \propto \ t^{-4}exp(-mL^2/2k_BTt^2). \qquad (2)$$

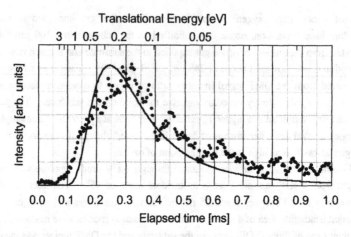

Figure 2 *QMF-TOF spectrum tuned for m/e = 48 (dots, O_3) at a T_s of 60 K and laser fluence of 13 mJcm^{-2}. The abscissa is the elapsed time after each ablation laser shot. The solid curve is the calculated TOF spectrum based on the Maxwell-Boltzmann distribution at 1,500 K.*

where, t is the elapsed time after the laser shot, m is the mass of an atom or molecule, L is the flight distance, k_B is Boltzmann's constant, and T is the temperature. The temperature of 1,500 K best explains the data, but the actual distribution is likely to approach equilibrium, showing a shoulder in the higher energy region and a tail in the lower energy region.

The QMF-TOF signals for O and O_2 were also observed for the same experimental conditions and their spectra are similar to the one shown in figure 2. They are best explained by the M-B distribution at 1,500 K, except for additional signals derived from decomposition of ozone at the ionizer. Oxygen molecules scattered on the chamber wall seem to have helped extend the tails to a lower energy below T_s, but ozone and the oxygen atoms on the wall would have reacted forming oxygen molecules. The signals for heavier species such as $(O_2)_n$ (n>1) were below the detection limit.

The dissociation energy of 98.3 kJmol^{-1} is equivalent to 11,800 K or 1.02 eV and is the unique origin of kinetic energy of the ablated species as there is little possibility of MPI at the present laser fluence. The total flux of the ablated species can only roughly be estimated to be $10^{15} \sim 10^{16}$ per laser shot. This is in the same range of the number of photons per shot (2.6×10^{15} for the same fluence). The fitted temperature of 1,500 K is only a fraction of 11,800 K, but it rises as the laser fluence increases as shown in figure 3 for T_s = 30 K. There must be several energy dispersion paths rather than a direct transfer to the translational energy of the ablated species. However, the ablation conditions, including the thickness of the solid ozone, must be optimized to get an ozone beam with a larger kinetic energy.

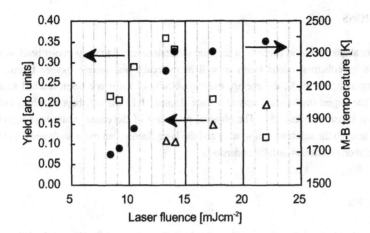

Figure 3 *Yield of ozone (squares), atomic oxygen (triangles), and fitted M-B temperature (dots) versus laser fluence for $T_s = 30$ K.*

The yield of ozone and oxygen atoms as a function of laser fluence at $T_s = 30$ K is plotted in figure 3 along with the M-B temperature fitted to the TOF spectra. The ozone yield is maximum at a fluence of 13 mJcm^{-2} at which the M-B temperature begins to saturate at 2,300 K. A similar dependence of the kinetic energy on laser fluence from 7 to 22 mJcm^{-2} was reported for Cl$_2$ [8], but the opposite behavior was observed for NO$_2$ [7]. The critical laser fluence of 13 mJcm^{-2} is probably the point at which the photon flux (2.6×10^{15}) exceeds the single-photon absorption limit for the total number of ozone molecules within the absorption depth, which is roughly estimated to be 10^{16}. Above the critical value, the multiphoton ionization of ozone molecules may partially occur, causing a weak plasma that enhances the decomposition of ozone and results in a drop of the ozone yield and a rise of the yield of atomic oxygen as shown in figure 3. There may not be an energy transfer from the weak plasma to the neutral ablated species as indicated by the saturation of the fitted temperature, but the production of ions must be checked for laser fluences above the critical value.

The kinetic energy of the ozone beam for the same laser fluence was somewhat larger for $T_s = 30$ K than for $T_s = 60$ K. A difference of 30 K in this temperature region may considerably change several factors such as the purity, density, and thickness of the solid ozone. For that reason, it should be possible to control the physical properties of ozone beams through the substrate temperature and laser fluence alone, but more detailed study is needed before the mechanism for transferring energy to the ablated species is fully understood.

CONCLUSIONS

UV-laser ablation of solid ozone on a sapphire substrate attached to a cryocooler head was found to produce a hyperthermal ozone beam as well as molecular and atomic oxygen beams. At a substrate temperature of 30 K, the average thermal velocity of the ozone beam reached as high as 2,300 K. The largest ozone yield was at a laser fluence of 13 mJcm^{-2}, beyond which enhanced decomposition of ozone occurred. The physical properties of the ozone beam were found to be controllable through the substrate temperature and the laser fluence, but more work needs to be done before the ablation mechanism is fully understood.

REFERENCES

1. P. Vanysek, Electrochemical series, *Handbook of Chemistry and Physics*, ed. D.R. Lide (CRC Press, 1996) pp. **8**:20-**8**:25.
2. T. Shimizu, F. Hirayama, K. Oka, H. Nonaka, M. Matsuda, and K. Arai, *Appl. Phys. Lett.* **64**, 1289 (1994).
3. D.K. Lathrop, S.E. Lussek, and R.A. Buhrman, *Appl. Phys. Lett.*, **51**, 1554 (1987).
4. S. Hosokawa and S. Ichimura, *Rev. Sci. Instrum.*, **62**, 1614 (1991).
5. K. Nakamura, A. Kurokawa, and S. Ichimrua, *Jpn. J. Appl. Phys.* **39** (2000) in press.
6. A. Kurokawa, K. Nakamura, S. Ichimura, and D. W. Moon, *Appl. Phys. Lett.*, **76**, 493 (2000).
7. S. Sekine, S. Ichimura, H. Shimizu, and H. Hashizume, *Jpn. J. Appl. Phys.*, **33**, L387 (1994).
8. L.M. Cousins and S.R. Leone, *Chem. Phys. Lett.*, **155**, 162 (1989).
9. M. Horvath, L. Bilitzky, and J. Huttner, *Ozone* (Elsevier, 1985) pp. 15.
10. H. Okabe, *Photochemistry of Small Molecules* (John Wiley & Sons, 1978) pp.237-244. The absorption coefficient therein of 120 atm^{-1}cm^{-1}, 0°C, base 10 and the density of 1.6 gcm^{-3} (page 22 of ref. 9) were used to estimate the absorption depth.
11. A.V. Bemderskii and C.A. Wight, *J. Chem. Phys.*, **101**, 292 (1994).
12. R.K. Sparks, L.R. Carlson, K. Shobatake, M.L. Kowalczyk, and Y.T. Lee, *J. Chem. Phys.*, **72**, 1401 (1980).

Mat. Res. Soc. Symp. Vol. 617 © 2000 Materials Research Society

Recent Progress on the Modeling of Laser Surface Cleaning

Y.F. Lu, W. D. Song, B.S. Lukyanchuk, M.H. Hong and W.Y. Zheng
Laser Microprocessing Laboratory, Department of Electrical Engineering and Data Storage Institute, National University of Singapore, 10 Kent Ridge Crescent, Singapore 119260,

ABSTRACT

Laser cleaning has emerged to effectively remove contaminants from solid surfaces. In this paper, recent progress on laser cleaning has been studied. First, a cleaning model is established for removal of particles from substrate surfaces. The model not only explains the influence of fluence on cleaning efficiency, but also predicts the cleaning thresholds. Following that, the optical resonance and near field effect are discussed for transparent particles with a size of $a \sim \lambda$ (radiation wavelength) which strongly influences the intensity distribution in the contacted area (substrate surface). The characterization of ejected particles during laser cleaning is finally investigated. It is found that the particle distribution curves closely fit to Gaussian curve.

INTRODUCTION

Recently, laser cleaning was demonstrated to be an efficient cleaning method for removal of particles from solid surfaces [1-21]. Two types of laser cleaning have been reported in the literature, relying on pulsed laser heating of the solid surfaces without or with the presence of a thin liquid coating. We shall refer to these two types as dry laser cleaning and steam laser cleaning, respectively. For dry laser cleaning, particles can be ejected from particulate-contaminated surfaces by short-pulse laser irradiation. The proposed mechanism of the ejection is fast thermal expansion of the particle and/or solid surfaces, which induces large cleaning force to overcome the adhesion force between particles and solid surfaces. For steam laser cleaning, the proposed mechanism is the momentum transfer from the laser-heated and suddenly evaporating liquid film to the particles on the solid surfaces. The following discussion will focus on recent progress on laser cleaning including cleaning model for laser-induced removal of particles from solid surfaces, the optical resonance and near field effect for transparent particles and characterization of ejected particles during laser cleaning.

CLEANING MODEL

Van der Waals force is one of the major forces between a particle and a rigid solid surface. If the shortest distance between the particle and the substrate surface is H_0 and the particle has a radius of R, the adhesion energy W_s in absence of deformation should be [22]

$$W_s(H_0) = \frac{AR}{6H_0} \qquad (1)$$

where A is the Hamaker constant which is determined by the material properties of both particle and substrate. H_0 is the separation distance between particle and solid surface.

Supposing that the particle is near the surface to a point where $H_0 = \varepsilon_0$ (~4 Å), the particle begins deforming due to the strong pressure on the contacting area. This status is called "point contact". Derjaguin and co-works [22-24] have investigated the deformation induced by the adhesion of particles. If no other load is exerted on particle, the initial deformation parameter Lp_0 is expressed as

$$Lp_0 = \frac{2^{\frac{1}{3}} R^{\frac{1}{3}} A^{\frac{2}{3}} (1 - \sigma^2)^{\frac{2}{3}}}{8 \varepsilon_0^{\frac{4}{3}} E^{\frac{2}{3}}} \qquad (2)$$

where σ and E are Pisson coefficient and Young's modulus of the particle, ε_0 is the least distance between the particle surface and the substrate. The elastic repelling force F_e and energy of the deformation W_e are

$$F_e = \frac{4R^{\frac{1}{2}} E}{3(1 - \sigma^2)} Lp_0^{\frac{3}{2}} \qquad (3)$$

and

$$W_e = \frac{8R^{\frac{1}{2}} E}{15(1 - \sigma^2)} Lp_0^{\frac{5}{2}} \qquad (4)$$

respectively.

Suppose that the substrate expands with constant velocity Ls/τ during the time of laser irradiation. (Ls is the maximum thermal expansion and τ is the laser pulse width). The particle dislocation is a function of time t, say $f(t)$. Particle deformation at time t can be expressed as

$$Lp(t) = \frac{Ls}{\tau} t - f(t) + Lp_0 \qquad (5)$$

The acceleration due to the elastic force can be expressed as:

$$\frac{4}{3} \pi R^3 \rho_p \frac{d^2 f(t)}{dt^2} = \frac{4R^{\frac{1}{2}} E}{3(1 - \sigma^2)} [Lp^{\frac{3}{2}}(t) - Lp_0^{\frac{3}{2}}] \qquad (6)$$

where ρ_p is the density of the particle. The initial conditions for Eq. 6 are

$$\frac{df}{dt}\Big|_{t=0} = 0 \qquad f\big|_{t=0} = 0 . \qquad (7)$$

Although this process happens only in a short time duration, it determines the kinetic energy and the elastic potential energy necessary to overcome the adhesion force. The particle deformation $Lp(t)$ and velocity $df(t)/dt$ can be calculated with the four-order Rounge-Kutta method [25].

Particle removal starts at time τ when substrate expansion reaches the maximum as pulse irradiation finishes. The particle ejection condition can be expressed as:

$$\frac{8R^2E}{15(1-\sigma^2)}[Ls-f(\tau)+Lp_0]^{\frac{5}{2}}+\frac{1}{2}\times\frac{4}{3}\pi R^3\rho(\frac{df}{dt}\mid\tau)^2 \geq F_s[Ls-f(\tau)+Lp_0]+\frac{AR}{6\varepsilon_0} \quad (8)$$

The first item in the left-hand side of the above equation is the elastic deformation potential energy after irradiation, the second item in the left side is the kinetic energy of the particle. The first item in the right side is the work done by adhesion force while the particle recovers from deformation from time instant τ to "point contact", the second item in the right side is the adhesion energy of "point contact".

F_S is the total surface adhesion force. As discussed by Derjaguin and Muller et al, F_S drops rapidly to a constant as deformation parameter $Lp(t)$ is greater than $0.1\varepsilon_0$ [22]. We take adhesion force in the process of substrate thermal expansion as

$$F_s = \frac{AR}{12\varepsilon_0^2} . \quad (9)$$

If we define ejecting energy Ej as the left part of Eq. 8 subtracting the right part, the particle removal condition can be expressed as

$$Ej > 0 \quad (10).$$

In the process of laser irradiation, substrate thermal expansion can be expressed as

$$Ls(t) = \int_0^\infty \gamma(T)\Delta T(x,t)dx , \quad (11)$$

$$\Delta T(x,t) = T(x,t) - T_0 \quad (12)$$

where γ is the linear thermal expansion coefficient of substrate material, $T(x,t)$ is the temperature at any point below the substrate surface, and T_0 is the initial temperature at the substrate surface. The temperature profile is governed by the one-dimensional heat equation

$$\rho_s(T)Cp(T)\frac{\partial(x,t)}{\partial t}=\frac{\partial}{\partial x}[K(T)\frac{\partial T(x,t)}{\partial x}]+(1-R_s)\alpha I_0\exp(-\alpha x) \quad (13)$$

where $\rho(T), Cp(T),\ K(T), R_s,\ \alpha$ and I_0 are density, specific heat, thermal conductivity, reflectivity, absorption coefficient of the substrate material and laser intensity on substrate surface, respectively.

If we know the thermal properties of the substrate materials, the surface thermal expansion in the process of laser irradiation can be calculated from Eqs. (11)- (13).

Figure 1 shows the theoretical results of particle ejection energies as a function of thermal expansion. If the ejection energy less than 0, the particle cannot be removed. It is shown that the ejection energy increases with increasing thermal expansion, which is almost in proportional to the laser fluence. Thus the particles are easier to be removed as laser fluence increases. This was confirmed by the experimental results in Fig. 2.

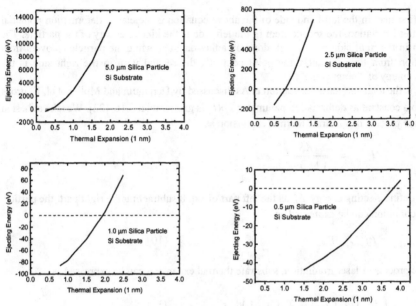

Fig. 1 The theoretical results of particle ejection energies as a function of thermal expansion.

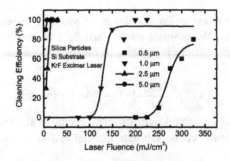

Fig. 2 Laser cleaning efficiency dependence with the laser fluence

From Fig. 2, it is found that the laser cleaning efficiency increases sharply along with laser fluence. For particles with a size of 0.5 and 1.0 μm, threshold laser fluences are about 100 and 225 mJ/cm^2, respectively. For larger particles with size of 2.5 and 5.0 μm, the threshold laser fluences are found to be lower than 5 mJ/cm^2.

Figure 3 shows the threshold laser fluences for particles with different size both theoretically and experimentally. For particle size bigger than 1 μm, such as 2.5 and 5 μm, the threshold laser fluences do not change obviously for different particle sizes. As particle size decreases to less than 1 μm, the threshold laser fluence increases sharply. This can explain why the submicron particles are more difficult to remove. Figure 3 also shows that the theoretical threshold laser fluences are greater than experimental results. The possible reason for the difference is that we assumed the incident light was not influenced by the particle. In fact, micro-focusing and light scattering may promote the light intensity near the particle-substrate contacting area since the particle is spherical. Following will discuss this effect due to micro-focusing and light scattering by a transparent spherical particle.

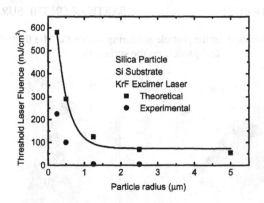

Fig. 3 The threshold laser fluences for particles with different size
both theoretically and experimentally

OPTICAL RESONANCE AND NEAR FIELD EFFECTS

The simplified preliminary examination of the intensity distribution can be done on the basis of the Mie theory. Under this approximation we ignore the secondary scattering effects, produced by radiation reflected by substrate surface (see in Fig. 4). The simplified consideration is useful as the first step, which permits to understand the role of the optical resonance effects in the near-field focusing.

The optical resonance is described by the Mie theory [26-30]. This theory presents the solution of Maxwell equations for the scattering of monochromatic electromagnetic wave with frequency by a sphere, dielectric permittivity, and magnetic permittivity holds for the complex refractive index.

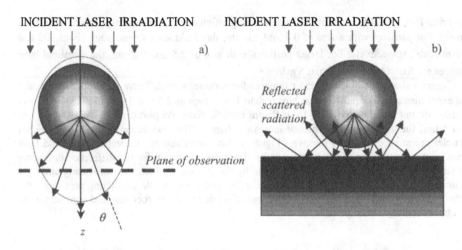

INCIDENT LASER IRRADIATION   INCIDENT LASER IRRADIATION

a)   b)

Reflected scattered radiation

Plane of observation

θ

z

SCATTERED RADIATION   PARTICLE ON THE SURFACE

Fig. 4. Schematic for the particle scattering within the Mie theory (a)
and particle on the surface (b)

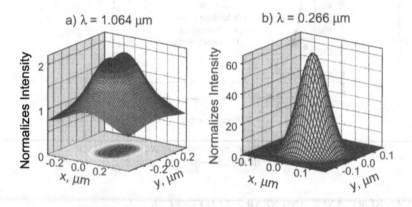

a) $\lambda = 1.064$ μm   b) $\lambda = 0.266$ μm

Fig. 5 The light intensity distributions on the surface
under the SiO_2 particle for two size parameters

To estimate the intensity of radiation, which will be absorbed by substrate surface, one can calculate from the Mie theory distribution of laser intensity in this region (see Fig. 4a). As an example in Fig. 5, the distributions are shown within the plane perpendicular to the wave vector of the incident wave. This intensity, I, is defined as absolute value of the z- component of time-averaged Poynting vector. The distributions are shown for SiO_2 sphere with two size parameters.

One can see that with IR-radiation $\lambda = 1.064$ μm and particle size $a = 0.25$ μm variations in the intensity over the surface are not very high, e.g. intensity in the center just twice higher than the incident intensity.

At the same time with UV-radiation $\lambda = 0.266$ μm and particle size $a = 0.5$ μm one can see a strong focusing effect, when intensity of light in the center is 60 times greater than the incident intensity, i.e. small particle works as a lens.

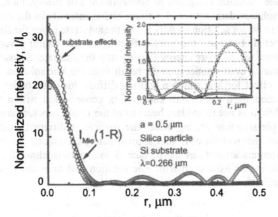

Fig. 6. Near-field light intensity along the radial axis with $\varphi = 45°$, calculated with the Mie theory (triangular) and from reflective matrix (circles).

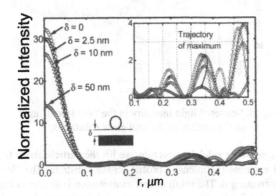

Fig. 7. Scattered light intensity on the substrate versus the distance between particle and the substrate.

The surface influenced the intensity distribution due to reflection and secondary scattering of the reflected radiation. It is not clear how the intensity distribution will vary due to this secondary scattering. In the limit of geometrical optics well-know effect, Newton rings, arise due to interference of scattered and reflected radiation.

We have to use in calculations the exact solution of the problem "particle on the surface" (in general case a particle in Fig. 4b is located at a distance δ above the substrate). This solution was found in [31]. Although this solution is rigorous, the idea of solution is rather simple.

The main results are presented in Figs. 6-8. These figures illustrate the main peculiarities of "particle on the surface" solution compared to conventional Mie theory. First, from Fig. 6 one can see that the true solution demonstrates 1.5 times higher intensity in the center as compared with Mie solution. Here we set solution slightly below the substrate to see the absorbed radiation. This result was expected, because the multi-reflecting of the Pointing vector between the particle and the substrate results more energy flowing into the substate. Second, the FWHM for the intensity distribution is even smaller than the result from the Mie solution. It means that the near-field sharpening effect for Mie solution will not diffuse due to secondary scattering. This result can be understood from the overall energy flux conservation within the range of the particle size. Meanwhile it is not so evident because of the ripple structure arising in near field region.Third, the ripple structure arising under the particle is different from the elementary Newton rings. It is true, that number of ripples approximately corresponds to the value. But also some additional maximums arise when the surface crosses zero-isoclines for Poynting vector field. Thus, these maximums exist even with Mie solution and they can be enhanced due to secondary scattering effects.

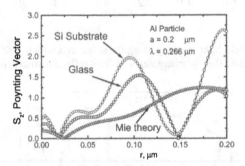

Fig. 8. Scattered light intensity in the "shadowed" area
on the surface of Si and glass substrates.

To analyze the origin of the sidelobe structure, we lift the particle above the substrate with a small distance δ, and examine the intensity profile on the substrate. The Newton rings should shrink inwards with increasing δ. The result of this examination is shown in Fig. 7. It shows that sidelobes firstly (up to the $\delta = 3$ nm) moves outwards with increasing δ, which implies that the ripples are mainly due to the reflecting of the near-field scattering light. Nevertheless with further increasing of δ they start to move inward. Thus we consider that this is comlex superposition of near-field Newton rings with Mie scattering. The noticible effect is the decreasing of maximal intensity in near field region. It enable optically measuring the shift δ by the order of 1 nm! It is significantly more sensitive than reasonable interferometric method.

To analyze the role of the particle and substrate materials we present in Fig. 8 the light intensity distribution in the "shadowed" area under the Aluminum particle, with two different substrates (Si and glass). The reflectivities of two subsrates are significantly different, about 0.7 and 0.08, respectively. It is found that the near field light intensity on Si is higher than that on glass surface. Similar effect can be responsible for the variation of the optical breakdown threshold of gas near the different surfaces [32].

CHARACTERIZATION OF EJECTED PARTICLES

Since the particle motion is determined by the cleaning process, characterisation of ejected particles may provide information to reveal the interactions among the laser, particles and the substrate. Therefore the investigation of particles after laser cleaning may lead to further understanding of the cleaning mechanism.

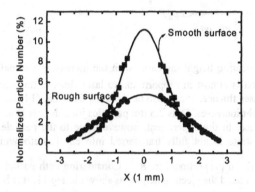

Fig.9 Particle distribution from a smooth surface and a rough surface.

Figure 9 shows the measured particle distribution after laser irradiation on a smooth and rough Si substrates. It is found that the particle distribution curves closely fit to Gaussion curve. From the results we can conclude that particles from the smooth surface produce a higher concentration in the centre than from the rough surface. The FWHM of the distribution curve of the rough surface is much wider than that of the smooth surface, which implies that particles from the rough surface were ejected more "randomly" in direction.

To measure the ejecting energy from a smooth surface, we set the height constant, then increased the laser fluence untill the ejected particles could reach the "capturing surface". In this way we can determine the fluence corresponding to the "ejecting energy". Figure 10 shows the particle ejection height as a function of incident laser fluence. From this figure we can conclude that the initial kinetic energy increases with incident laser fluence. It is also observed that 2.5 μm particles can "jump" higher than 5.0 μm particles under the same laser fluence. Ignoring the influence of adhesion force, the ejecting height can be expressed as[20],

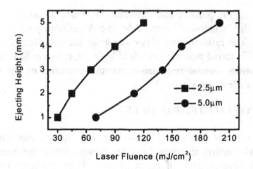

Fig.10 Particle Ejection height in dependence of laser fluence.

$$h = \frac{W_e}{\frac{4}{3}\pi R^3 \rho g} \propto L p_0^{\frac{5}{2}} R^{-\frac{5}{2}}. \qquad (14)$$

we can conclude that the ejecting height increases with the increasing thermal expansion Lp_0. Since the thermal expansion is almost in proportional to laser fluence, the ejecting height will increase with increasing laser fluence. It can also be deduced from Eq. (14) that as particle size decreases, the ejecting height increases. Thus 2.5 μm particles jump higher than 5.0 μm particles under the same laser fluence. In the experiment, however, due to the particle movement, the thermal expansion of substrate is not fully transferred into elastic deformation, so that the ejecting height is not strictly in proportional to $R^{-\frac{5}{2}}$. Considering both particle movement and elastic deformation, we calculated the ejecting height, as shown in Fig.11. It is in agreement with the experimental results.

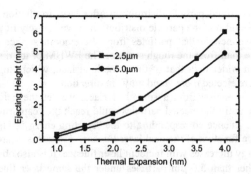

Fig.11 Calculated particle eject ion height in dependence of thermal expansion.

CONCLUSIONS

In summary, laser cleaning was demonstrated to be an efficient cleaning tool for removing contaminants from solid surfaces both theoretically and experimentally. Cleaning model was established for laser-induced removal of particles from solid surfaces. The model not only explains the influence of fluence on cleaning efficiency, but also predicts the cleaning thresholds. The optical resonance and near field effect were examined in the paper. Results of calculations demonstrate that small transparent particles play a role of focusing lens, producing high intensities on the surface under the particle. The additional ripples arising around the main peak present a comlex superposition of near-field Newton rings effect with Mie scattering. We consider that these ripples may play an important role in laser cleaning with short laser pulses. The characterization of ejected particles during laser cleaning is also investigated. The angular distribution of ejected particles fits to a Gaussian curve. The particles from a smooth substrate mostly concentrated on the center of the capturing surface, while those from a rough surface were ejected more randomly in direction.

REFERENCES

1. W. Zapka, W. Ziemlich, and A.C. Tam, Appl. Phys. Lett., 58, 2217 (1991).
2. K. Imen, S.J. Lee and S.D. Allen, Appl. Phys. Lett., 58(2), 203(1991).
3. A.C. Tam, W.P. Leung, W. Zapka and W. Ziemlich, J. Appl. Phys. 71(7), 3515(1992).
4. J.D. Kelley and F.E. Hovis, Microelectronic Engineering 20, 159(1993).
5. Y.F. Lu, M. Takai, S. Komuro, T. Shiokawa and Y. Aoyagi, Appl. Phys. A 59, 281(1994).
6. H.K. Park, C.P. Grigoropoulos, W. P. Leung and A.C. Tam, IEEE Transactions on Components, Packaging, and Manufacturing Technology Part A, 17(4) 631(1994).
7. Y.F. Lu, W.D. Song, M.H. Hong, B.S. Teo, T.C. Chong and T.S. Low, J. Appl. Phys., 80(1), 499(1996).
8. K. Mann, B. Wolff-Rottke and F. Muller, Appl. Surf. Sci., 98, 463(1996).
9. R. Oltra, O. Yavas, F. Cruz, J.P. Boquillon and C. Sartori, Appl. Surf. Sci., 98, 484(1996).
10. D.A. Wesner, M. Mertin, F. Lupp and E.W. Kreutz, Appl. Surf. Sci., 98, 479(1996).
11. M. Afif, J.P. Girardeau-Montaut, C. Tomas, M.Romamd, M. Charbonnier, N.S. Prakash, A. Perez, G. Marest and J.M. Frigerio, Appl. Surf. Sci., 98, 469(1996).
12. I. Gobernado-Mitre, J. Medina, B. Calvo, A.C. Prieto, L.A. Leal, B. Perez, F. Marcos and A.M. de Frutos, Appl. Surf. Sci., 98, 474(1996).
13. J.B. Heroux, S. Boughaba, I. Ressejac, E. Sacher and M. Meunier, J. Appl. Phys. 79(6), 2857(1996).
14. W.D.Song, Y.F.Lu, K.D.Ye, C.K. Tee, M.H. Hong, D.M. Liu and T.S.Low, SPIE Vol.3184, 158(1997).
15. Y.F.Lu, W.D.Song, B.W. Ang, M.H.Hong, D.S.H.Chan and T.S.Low, Appl. Phys. A65, 9(1997).
16. Y.F.Lu, W.D.Song, K.D.Ye, Y.P.Lee, D.S.H.Chan and T.S.Low, Jpn. J. Appl. Phys., 36(10A), L1304(1997).
17. Y.F.Lu, W.D.Song, K.D. Ye, M.H.Hong, D.M. Liu, D.S.H.Chan and T.S.Low, Appl. Surf. Sci. 120, 317(1997).

18. Y.F.Lu, W.D.Song, C.K. Tee, D.S.H.Chan and T.S.Low, Jpn. J. Appl. Phys. Part 1, No.3A, 840(1998).
19. Y.F.Lu, Y.W. Zheng, W.D.Song, Appl. Phys. A68, 569(1999).
20. Y.F.Lu, Y.W. Zheng, W.D.Song, J. Appl. Phys. 87(3), 1534(2000).
21. Y.F.Lu, Y.W. Zheng, W.D.Song, J. Appl. Phys. 87(1), 549(2000).
22. B.V. Derjaguin, V.M. Muller and YU.P. Toporov, Journal of Colloid and Interface Science, 53(2), 314(1975).
23. V.M Muller, V.S. Yushchenko and B.V. Derjaguin, Journal of Colloid and Interface Science, 77(1), 91(1980).
24. V.M. Muller, V.S.Yushchenko and B.V. Derjaguin, Journal of Colloid and Interface Science, 92(1), 92(1983).
25. W.H. Press, S.A. Teulolsky, W.T. Wetterling and B.P. Flannery, *Numerical Recipes in C, The Art of Scientific Computing*, Second Edition, Cambridge University Press, 1992.
26. J. A. Stratton: *Electromagnetic Theory*, (McGraw-Hill, New York & London, 1941).
27. H. C. Van de Hulst: *Light Scattering by Small Particles*, (Dower Publ., New York, 1981).
28. C. E. Bohren, D. R. Huffman: *Absorption and Scattering of Light by Small Particles*, (John Willey & Sons, 1983).
29. M. Born, E. Wolf: *Principles of Optics*, 7-th Edition, (Cambridge University Press, 1999).
30. M. Kerker: *The scattering of Light*, (Academic Press, New York & London, 1969).
31. P. A. Bobbert, J. Vlieger, Physica A, vol. **137**, pp. 209-242 (1986).
32. A. M. Prokhorov, V. I. Konov, I. Ursu, I. Mihailescu: *"Interaction of Laser Radiation with Metals"* (Moscow, Nauka, 1988).

Mat. Res. Soc. Symp. Vol. 617 © 2000 Materials Research Society

Localized Excimer Laser Energy Modulation in the Crystallization of poly-Si Film on Stepped Substrate

Kee-Chan Park, Sang-Hoon Jung, Woo-Jin Nam, and Min-Koo Han
School of Electrical Engineering, Seoul Nat'l Univ., Seoul, 151-742, KOREA.
Tel. +82-2-880-7992, Fax. +82-2-883-0827, E-mail:pkch@emlab.snu.ac.kr

ABSTRACT

A new excimer laser recrystallization method of a-Si film to increase the grain size of poly-Si film has been proposed. Excimer laser energy was locally modulated by being irradiated on stepped substrate with 500 nm deep trench on which a-Si film was deposited. Fairly large poly-Si grains (over 1 μm) were obtained due to lateral thermal gradient which resulted from the laser energy difference on the vertical wall and on the horizontal bottom plane of the trench without altering laser energy density elaborately.

INTRODUCTION

The polycrystalline silicon thin film transistor (poly-Si TFT) fabricated employing the excimer laser annealing (ELA) of amorphous silicon (a-Si) film is considered a promising device for active matrix liquid crystal display (AMLCD) due to high current driving capability [1]. Poly-Si film recrystallized by the cnventional ELA usually result in rather small grain size that is typically less than 1 μm so that the device performance of poly-Si TFT is not satisfactory compared with single crystalline MOS transistors [2].

The purpose of our work is to report a new laser recrystallization method of a-Si film employing the eximer laser energy modulation on the substrate. Large poly-Si grains exceeding 1 μm were successfully formed and a-Si channel offset region was also formed by employing the proposed method. The advantage of our proposed method over the conventional ELA is the feasibility of large grain growth that is less depndent on the laser energy variation and the possible application of the residual a-Si region as a poly-Si channel offset to reduce the large off current of poly-Si TFTs [3].

EXPERIMENT

Figure 1 shows the TFT structure to implement the proposed laser energy modulation method. a-Si active layer was deposited on a 0.5 μm deep trench-formed substrate and recrystallized by ELA at a laser energy density of 350 mJ/cm² with 100 nm or 300 nm thick capping SiO$_2$. Due to the laser energy difference between on the horizontal and on the vertical trench wall, the laser-irradiated silicon film has lateral temperatrue gradient and silcon grains are motivated to grow laterally from the relatively low-temperatured bottom corner of the trench wall to high-temperatured liquid silicon on the bottom plane of the trench as shown in figure 2 [4].

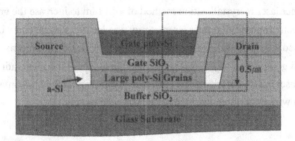

Figure 1. *The structure of the proposed poly-Si TFT. The square area enclosed by dotted line is redrawn in figure 2.*

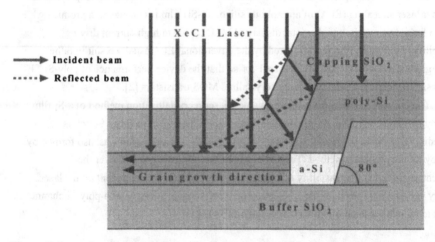

Figure 2. *Lateral grain growth due to the laser energy modulation on vertical trench wall.*

On the trench wall, two reflections take place on the air-to-oxide interface and the oxide-to-silicon interface respectively. Due to the double reflection effect, the laser energy density of oblique incident beam become more weak than when the capping oxide is not covered.

RESULTS

Figure 3 shows the trasmission electron microscopy (TEM) images of the recrystallized poly-Si active layer employing the proposed ELA method. Very large poly-Si grains (see figure 3) compared with conventional case (see figure 4) were grown in the trench bottom plane by suppressing undesirable radom necleation. The growth of uniformly large grains is due to the temperature gradient in the molten silicon layer which resulted from the local laser energy modulation. a-Si region remained at the bottom corner of the trench after the laser recrystallization and the a-Si region can be utilized as an offset between poly-Si channel and drain of TFT in order to decrease the large off-current of poly-Si TFT [3].

Figure 5 shows the TEM images of the recrystallized poly-Si active layer with 100 nm thick capping oxide and without capping oxide. Small poly-Si grains (see figure 5) compared with those with 300 nm thick capping oxide (see figure 3) were formed in the trench bottom and very long a-Si region was formed at the bottom corner of the trench wall. This is attributed to the rather low heat conservation capability of the thin capping oxide. Therefore it can be concluded that the heat conservation during the poly-Si grain growth is also important as well as the induction of the lateral thermal gradient.

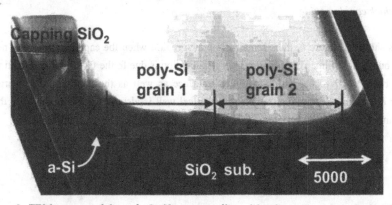

Figure 3. TEM images of the poly-Si film recrystallizaed by the proposed method with 300 nm thick capping oxide.

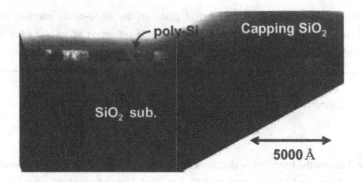

Figure 4. TEM images of the poly-Si film recrystallized by the conventional ELA method with 300 nm thick capping oxide.

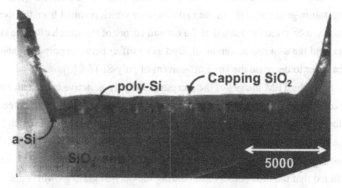

Figure 5.   TEM images of the poly-Si film recrystallizaed by the proposed method with 100 nm thick capping oxide.

As shown in figure 6, there remained no a-Si region when the capping oxide was not covered on the a-Si film. This complete recrytstallization is due to the fact that the reflection of laser beam on the steep oxide-to-silicon interface as well as the incident laser energy decrease due to the oblique incidence at the vertical wall is also important to modulate the laser beam energy on the substrate.

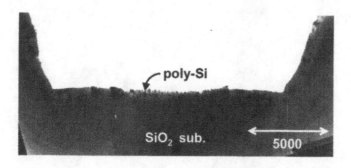

Figure 6. *TEM images of the poly-Si film recrystallized by the proposed method without capping oxide.*

CONCLUSION

We have proposed and investigated the excimer laser energy modulation effect on the substrate. Due to the laser energy difference on the vertical wall and horizontal bottom plane of the trench in the substrate, very large (over 1 μm) poly-Si grains were formed. Different recrystallization results for various capping oxide thickness' were also investigated and discussed. In addition, residual a-Si region was preserved in the bottom corner of the trench wall and it can be utilized as a channel offset of poly-Si TFT to reduce the large off current of poly-Si TFT.

REFERENCES

1. J. Ohwada et al., *SSDM Ext. Abs.*, 55 (1987).
2. H. Kuriyama, S. Kiyama, S. Noguchi, T.Kuwahara, S. Ishida, T. Nohda, K. Sano, H. Iwata, H. Kawata, M. Osumi, S. Tsuda, S. Nakano and Y. Kuwano, *Jpn. J. Appl. Phys.*, **30**, 3700 (1991).
3. K.H. Lee, W.Y. Lim, J.K. Park, J. Jang, *SID '97 Tech. Digest*, 481 (1997).
4. J.H. Jeon, M.C. Lee, K.C. Park and M.K. Han, *Jpn. J. Appl. Phys.*, **39**, 101 (2000).

Mat. Res. Soc. Symp. Vol. 617 © 2000 Materials Research Society

Simulation of Polycrystalline Silicon Growth by Pulsed Excimer Laser Annealing

Toshio Kudo, Daiji Ichishima and Cheng-Guo Jin*
Research & Development Center, Sumitomo Heavy Industries Ltd.,
63-30 Yuuhigaoka, Hiratsuka, Kanagawa 254-0806, JAPAN
Tso_Kudo@shi.co.jp
*ACT Center, TIC Corporation,
2-9-30 Kitasaiwai, Nishiku, Yokohama, Kanagawa 220-0004, JAPAN

ABSTRACT

The dynamic simulation of poly-Si film synthesis has been fulfilled by means of the over-lapping irradiation of single- and double-pulsed XeCl excimer lasers shaped into the line beam. A novel model applied to the dynamic simulation is based on the homogeneous nucleation, and the growth and shrinkage velocity of Si grains. The results simulated with the single-pulsed XeCl excimer laser has reproduced the super lateral growth (SLG) phenomenon which occurs in the very narrow range of energy density (the near complete melt regime). The actual energy density dispersion within 5.3% allows us to visualize the multiformity of grain sizes in the cross sectional texture. Standing on the reproduction of the SLG phenomenon by the single-pulsed irradiation, we have obtained the practical knowledge of the growth process of larger grains by the double-pulsed irradiation. We intend that the first pulse has charge of completely melting and the second pulse has charge of adjusting the number of nuclei. The adjustment of the first pulse energy density rather than that of the second pulse energy density leads to the growth of huge columnar grains much larger than the thickness of the Si layer. For the double-pulsed irradiation, the influence of the worst energy density dispersion (each of double pulses has the same fluctuation as the single pulse) is much larger than that of the actual delay time jitter (within 2.5ns) in a sense of the multiformity of grain sizes.

INTRODUCTION

Low temperature polycrystalline Si (p-Si) technology has potential to realize System On Panel with not only built-in driver but also microprocessor, memory, sensor etc. on a glass substrate [1]. Consequently, it is no exaggeration to say that the pulsed-laser crystallization of amorphous Si (a-Si) films on the glass substrate is the most promising technology of fabricating poly-Si TFTs. But there still now remain improvements in the quality of p-Si films to advance the performance of poly-Si TFTs: the enlargement and uniformity of Si grain size, strong crystal orientation and smooth surface morphology [2-4]. From the need to control the polycrystallization processes, the visualized computer simulation of p-Si film synthesis should be a powerful tool to predict the polycrystallization process under the repetition of pulsed-laser irradiation. The dynamic simulation of p-Si film synthesis, however, had not yet been done in the poly-Si TFT technology.

Ichishima succeeded in the dynamic simulation of p-Si film synthesis [5]. A novel model of simulation is characterized by the growing and shrinking of Si grains. We report on the simulation results of the p-Si film synthesis by means of the overlapping irradiation of single- and double-pulsed excimer lasers.

A SIMULATION MODEL

We introduce the novel simulation model proposed by Ichishima [5]. The growth and shrinkage of Si grains are essential to the crystallization process of a-Si films by the pulsed excimer laser annealing, because p-Si films are synthesized in the repetition of melt and growth by the overlapping irradiation of a line beam. This model is based on the homogeneous nucleation, and the growth and shrinkage velocity of Si grains [6, 7].

The 2D numerical calculation is described as follows. The a-Si layer 1μm wide and 500Å thick, which has a SiO$_2$ underlayer (1000Å thickness) on a glass substrate at room temperature, is melted and crystallized in vacuum by a pulsed XeCl excimer laser (λ: 308nm). The pulse shape is formed into a Gaussian profile and the pulse duration is 14ns, defined by the full width at half maximum (FWHM). The beam shape is a line beam whose spatial intensity distribution is given by the Gaussian profile (FWHM: 0.6mm), and the line beam is irradiated at the overlap ratio of 90% (a step of scanning: 0.06mm). The temperature distributions induced by the absorption of the laser light within the a-Si layer are calculated on the basis of the 2D heat equation [8]

$$\rho C \frac{\partial T}{\partial t} = \nabla [\kappa(T) \nabla T] + \dot{Q} + \rho L \dot{\xi},$$ (1)

where ρ is the mass density, C the specific heat, κ the thermal conductivity, Q the heat source, L the latent heat and ξ the crystal phase fraction. The dot (•) symbolizes the differential with respect to time. The heat source term in Eqn. (1) can be written in the form

$$\dot{Q} = (1-R) P(t) \alpha \ exp(-\alpha z) f(x, z).$$ (2)

Here R is the optical reflectivity, $P(t)$ the laser power, α the optical absorption coefficient, and $f(x, z)$ the beam shape function. The line beam with a Gaussian shape function is scanned along the x-axis. The probability of nucleation is evaluated by

$$I(T(t)) = I_0 exp(-\frac{\Delta G}{k_B T}),$$ (3)

where ΔG is the free energy of a critical nucleus at T [6]. The probability (i.e. Eqn. 3) is obtained at each position within the Si layer through the temperature field. Applying the acceptance-rejection method, the nuclei are located at each position. The growth and shrinkage velocity of Si grains is obtained with the expression [7]

$$V(T) = V_0 p(T) (1 - exp(-\frac{\Delta G}{k_B T})).$$ (4)

And the crystal phase fraction can be calculated from the equation of continuity

$$\frac{\partial S}{\partial t} + \nabla(S V) = 0; S = 1 - \xi.$$ (5)

Carrying on the above-mentioned steps repeatedly with the progress of time, we can obtain the knowledge on the location, shape and size of Si grains within the cross sectional texture of the p-Si layers [8].

SINGLE PULSE IRRADIATION

Figure 1 shows, for the single pulse irradiation of the XeCl excimer laser under the

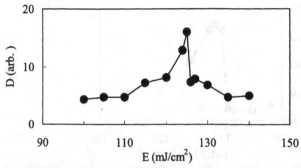

Figure 1. Plot of average Si grain size versus energy density for p-Si layers simulated by the single-pulsed irradiation.

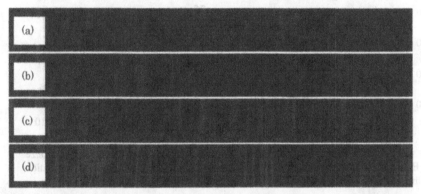

Figure 2. Cross sectional textures of p-Si layers simulated by the single-pulsed irradiation at the following energy density: (a) $105 mJ/cm^2$, (b) $124 mJ/cm^2$, (c) $130 mJ/cm^2$, (d) $124 mJ/cm^2$ (with energy dispersion).

conditions mentioned above, average Si grain size (D) versus beam energy density (E), where the value D is converted into the average diameter of circles equal to the areas of grains within the cross sectional view of the Si layer. A sharp increase in D occurs in the very narrow range of E [9]. Figures 2(a)-(c) are the visualized cross sections of p-Si layers at E = $105 mJ/cm^2$, 124 mJ/cm^2 and $130 mJ/cm^2$, each of which belongs to the partial melt regime, the near complete melt regime and the complete melt regime [10]. The cross sectional view of the p-Si layer in Figure 2(d) shows the influence of energy density dispersion at the sharp peak shown in Fig. 1. The dispersion data within 5.3% were actually measured with the XeCl excimer laser. The dispersion results in the reduction of the average grain size; the multiformity in grain sizes.

DOUBLE PULSE IRRADIATION

The typical double-pulse irradiation consists of the first pulse with the high energy density (E_1 = 130 mJ/cm^2) and the second pulse delayed with the low energy density (E_2 = 60 mJ/cm^2).

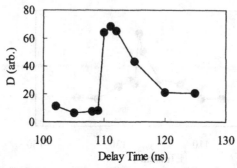

Figure 3. Plot of average Si grain size versus delay time for p-Si layers simulated by the double-pulsed irradiation ($E_1 = 130mJ/cm^2$, $E_2 = 60$ mJ/cm^2).

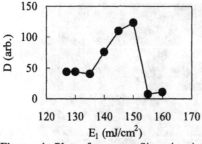

Figure 4. Plot of average Si grain size versus E_1 for the double-pulsed irradiation ($E_2 = 60mJ/cm^2$, delay time = 115ns).

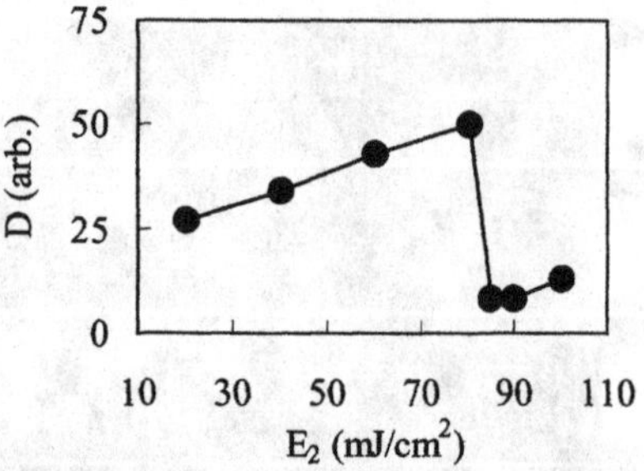

Figure 5. Plot of average Si grain size versus E_2 for the double-pulsed irradiation ($E_1 = 130mJ/cm^2$, delay time = 115ns).

The same XeCl excimer lasers mentioned above are used as the two light sources. We intend that the first pulse has charge of completely melting and the second pulse has charge of adjusting the number of nuclei. Figure 3 shows average grain size (D) versus delay time (τ), where the value τ is the separation in time between the first pulse and the second pulse. There appears a peak (FWHM: 7.5ns) around 111 ns. Figure 4 shows the effects of E_1 on D under $E_2 = 60mJ/cm^2$ ($\tau = 115ns$) and Figure 5 the effects of E_2 on D under $E_1 = 130mJ/cm^2$ ($\tau = 115ns$). The effects of E_2 are not as sensitive as the effects of E_1. Figures 6(a)-(c) show the visualized cross sections of p-Si layers for $\tau = 105ns$, 111ns and 120ns under the fixed energy densities, $E_1 = 130mJ/cm^2$ and $E_2 = 60mJ/cm^2$. Figure 6(d) is a cross sectional texture at a peak of the D vs. E_1 ($E_1 = 150mJ/cm^2$, $E_2 = 60mJ/cm^2$). The occurrence of a peak in the D vs. τ (Fig. 3) is similar to that in the D vs. E (Fig. 1).

The cross sectional views of the p-Si layers in Figures 6(e)-(f) show the influence of energy density dispersion and delay time jitter at the peak in Fig. 3 ($\tau = 111ns$). The worst dispersion of E for the double pulses is set: each of double pulses has the same fluctuation as the actual measurement (within 5.3%) for the single pulse. And the jitter within 2.5ns was actually measured with the same XeCl excimer laser. The energy density dispersion results in the reduction of the average grain size; the multiformity in grain sizes, but the delay time jitter has a small influence: keeping the grains columnar.

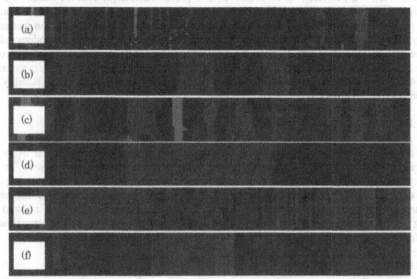

Figure 6. Cross sectional textures of p-Si layers simulated by the double-pulsed irradiation at the following delay time: (a) 105ns, (b) 111ns, (c) 120ns, (d) 115ns (E_1= 150mJ/cm^2, E_2= 60mJ/cm^2), (e) 111ns (with energy dispersion), (f) 111ns (with delay time jitter).

DISCUSSION

The cross sectional textures shown in Fig. 2 reveal the crystallization process of a-Si layers by the overlap irradiation of the single-pulsed XeCl laser. The Fig. 2(a) in the partial melting regime shows up a double-layer structure, an upper layer (coarse grains) and a lower layer (fine grains). The fine grains (seeds) are explosively produced when the a-Si layer is fused and super-cooled by the latent heat of solidification [11]. The coarse grains grow up at the liquid-solid interface [10]; the interface moves down and up with the step-scanning irradiation of the line beam formed into the Gaussian profile and the melt and growth is repeated. The Fig. 2(b) in the near complete melt regime shows the coexistence of large columnar grains and microctystals, caused by dense and less dense seeds at the interface of liquid-Si/solid-SiO$_2$, where the large grains result from the super lateral growth (SLG) phenomenon. And also the SLG phenomenon appears as a sharp peak in Fig. 1 [9]. The Fig. 2(c) in the complete melt regime is almost covered by microcrystals grown from the dense seeds at the bottom of the Si layer. The dispersion in E (within 5.3%) brings on the multiformity of grain sizes shown in Fig. 2(d): the coexistence of many microcrystals from the irradiation over the threshold value (E > 124 mJ/cm^2), the columnar grains from the SLG, and the irregular grains from seeds at the liquid-solid interface in the middle of the step-scanning irradiation.

The simulation results of the double pulse irradiation are classified into three regimes of texture: the multiformity of grain sizes in $\tau < 110$ns, the decay of columnar grains in $\tau > 120$ns and the growth of columnar grains in the middle range of τ. The comparison between Fig. 4

and Fig. 5 indicates that the severe control of E_1 is much more necessary than that of E_2. By the adjustment of E_1, it is possible to grow huge columnar grains much larger than the thickness of the Si layer as shown in Fig. 6(d) [12]. Under the performance of the XeCl excimer laser at present, the influence of the energy density dispersion (within 5.3%) is much larger than that of the delay time jitter (within 2.5ns) in a sense of the multiformity of grain sizes. To understand better the growth process of larger grains by the double-pulsed laser irradiation, however, we need the accumulation of knowledge about the interaction between the energy density dispersion and the delay time jitter, the influence of the laser beam shape, the effect of pointing stability, and so on through the proposed computer simulation.

CONCLUSIONS

We have fulfilled the computer simulation of the p-Si film synthesis by means of the overlapping irradiation of single- and double-pulsed XeCl lasers with the line beam shape. The simulation model is based on the homogeneous nucleation, and the growth and shrinkage velocity of Si grains. The results of the single-pulsed laser irradiation has demonstrated that the SLG phenomenon occurs in the very narrow range of E (the near complete melt regime). Building on the reproduction of the single-pulsed irradiation, we have obtained the practical knowledge of the growth process of larger grains by the use of the double-pulsed irradiation.

REFERENCES

1. Y. Funaki, NIKKEI MICRODEVICES, **2**, 160 (1998) (in Japanese).
2. R. Ishihara and M. Matsumura, IEEE Electron Dev. Lett. **31**, 1956 (1995).
3. H. Nishitani et al., IDW'98, ALCp-1 (1998)
4. K. Yoneda, Mat. Res. Soc. Symp. Proc. **507**, 47 (1998).
5. D. Ichishima et al. (unpublished).
6. D. H. Lowndes and R. F. Wood, Journal of Luminescence, **30**, 395 (1985).
7. S. R. Stiffler, P. V. Evans and A. L. Greer, Acta Metall. Mater., **40**, 1617 (1992).
8. P. K. Galenko and V. A. Zhuravlev, Physics of Dendrites (World Scientific, 1994) p.114.
9. J. S. Im, H. J. Kim and M. O. Thompson, Appl. Phys. Lett. **63**, 1969 (1993).
10. H. Watanabe, H. Miki, S. Sugai, K. Kawasaki and T. Kioka, Jpn. J. Appl. Phys. **33**, 4491 (1994).
11. M. O.Thomson, G. L. Galvin, J. W. Mayer, P. S. Peercy, J. M. Poate, D. C. Jacobson, A. G. Gullis and N. G. Chew, Phys. Rev. Lett. **52**, 2360 (1984).
12. B. Z. Bachrach, K. Winer, J. B. Boyce, S. E. Ready, R. I. Johnson, and G. B. Anderson, J. Electron. Mater. **19**, 241 (1990).

Mat. Res. Soc. Symp. Vol. 617 © 2000 Materials Research Society

Laser-Induced Liftoff and Laser Patterning of Large Free-Standing GaN Substrates

O. Ambacher[1], M. K. Kelly[2], C. R. Miskys, L. Höppel, C. Nebel, and M. Stutzmann

Walter Schottky Institute, Technical University of Munich
Am Coulombwall, D-85748 Garching, Germany

ABSTRACT

Free-standing GaN crystals are produced from 200-300 µm thick GaN films grown on 2 inch sapphire substrates by hydride vapor phase epitaxy. The GaN films are separated from the growth substrate by laser-induced liftoff, using a pulsed laser to thermally decompose a thin layer of GaN close to the film-substrate interface. The free-standing films are polished and used for the homoepitaxial growth of high quality GaN layers by metalorganic chemical vapor deposition. The structural and optical properties of the homoepitaxial films in comparison to layers grown on sapphire are significantly improved, mainly because of lower dislocation density and surface roughness as low as $5x10^6$ cm^2 and 0.2 nm, respectively.
Laser-induced thermal decomposition is also applied to achieve etching of GaN. At exposures of 500 mJ/cm^2 with 355 nm light, etch rates of up to 90 nm for one pulse are obtained. Illumination with an interference grating is used to produce trenches as narrow as 100 nm or sinusoidal surface patterns with a period of 260 nm. Such surface morphologies are very useful for the processing of anti-reflection coatings or distributed Bragg reflectors.

INTRODUCTION

To fully explore the large application potential of group-III-nitrides, it is a big advantage to achieve epitaxial growth on lattice-matched substrates. Although relatively small bulk GaN crystals can be reproducibly grown [1], large bulk single crystals of GaN or AlN for the fabrication of epi-ready substrates are not yet commercially available. Therefore all GaN-based devices realized up to date are deposited by heteroepitaxy, mostly on sapphire (Al$_2$O$_3$) or silicon carbide (SiC) substrates. Al$_2$O$_3$ features a relatively low cost and is the most commonly used substrate material for optoelectronic devices because of its large bandgap, thermal stability and epi-ready surface quality [2-4]. SiC is of special interest for high power devices like field effect transistors, hetero-bipolar devices and microwave amplifiers because of its high thermal

---

[1] *e-mail: ambacher@wsi.tu-muenchen.de,* [2] *mkelly@planet-interkom.de*

conductivity and the easy implementation of back contacts on n- or p-type doped substrates. However, the mismatch in lattice constants and thermal expansion coefficients between epitaxial layers of group-III-nitrides and the available substrates gives rise to a high density of dislocations (between 10^8 and 10^{10} cm^{-2}) and strain, limiting e.g the electron mobility, doping efficiency and lifetime of devices [5, 6].

The capability to produce superior material quality by homoepitaxial growth of GaN on GaN bulk crystals or bulk-like substrates has been demonstrated before [7]. At present, the largest GaN crystals are grown using the high-pressure method [8]. Such crystals (flat hexagonal plates) grow up to 100 mm^2 within 200 h, possess a very low dislocation density (10^4 cm^{-2}) and are highly n-type conductive due to the presence of oxygen impurities. GaN homoepitaxial layers grown by MOCVD on these crystals were described in a number of articles [9, 10]. Two-dimensional step-flow growth of GaN can be achieved without the need for additional deposition steps like surface nitridation, low temperature nucleation, or the growth of thick buffer layers. In x-ray diffraction, the (0004) reflexes of the substrate and of the MOCVD-grown layer could still be separated due to the differences between the larger lattice constants of the bulk GaN substrate (increased by point defects). Thus the lattice match was not perfect, both reflexes had rocking curves with FWHM of about 20 arcsec. The photoluminescence spectrum was dominated by the donor-bound exciton recombination with a FWHM below 1 meV (T = 5 K). Thus the structural and optical properties of homoepitaxial GaN are significantly improved in comparison to heteroepitaxial layers, but the size and the oxygen impurities of the high pressure grown substrates still limit the quality of lattice matching between epitaxial layer and substrate, the possibility to achieve p-type doping, and the scaling up for device fabrication.

An alternative to bulk crystals are bulk-like thick-film GaN substrates produced by hydride vapor-phase epitaxy (HVPE) [11, 12]. HVPE is a promising technique for the growth of high quality GaN at very high deposition rates (100 μm/h) and at reasonable cost. Recent developments in this technique have lead to a significant improvement of the structural quality and to the realization of large-area crack-free films (up to 2 inches in diameter and 300 μm thickness). Often the surface roughness is in the range of several microns and surface polishing is needed to prepare epi-ready GaN substrates. Polishing procedures are hampered by a bowing of the GaN/Al$_2$O$_3$ structures, caused by the differences in thermal expansion coefficients between GaN and sapphire leading to the accumulation of thermal stress during the cool down from the high growth temperature ($\approx$ 1050°C). In order to solve problems like bowing due to thermal stress or to facilitate thermal and electrical backside contacts, however, it is still desirable to remove the Al$_2$O$_3$ substrate. As an example, Nakamura et al. obtained improvements in the lifetime of laser diodes and cleaved facets after polishing the sapphire away [13].

In this paper we report the production of 2 inch free-standing GaN wafers by laser-induced liftoff, using a pulsed laser to thermally decompose a thin layer of GaN at the film-sapphire interface. Sequentially scanned pulses are employed to perform the liftoff using elevated temperatures to relieve postgrowth bowing. After polishing, the wafers are used for the homoepitaxy of GaN by MOCVD. Even if the homoepitaxy of GaN with high structural quality can be achieved, a variety of GaN applications will require a range of options for patterning and etching. Since GaN strongly resists wet etches, more complex, poisonous and expensive processes have been required, primarily those using Cl-based reactive ion etching (RIE) [14]. Here we report on etching of GaN films which is achieved by laser-induced thermal decomposition. In order to determine the achievable maximum etch rate, the decomposition of GaN films is measured versus intensity of the laser pulses. A more advanced technique of patterning by pulsed laser holography is also presented, which is used to realize a sine wave surface pattern with periods suitable for processing of antireflection coatings or distributed Bragg reflectors.

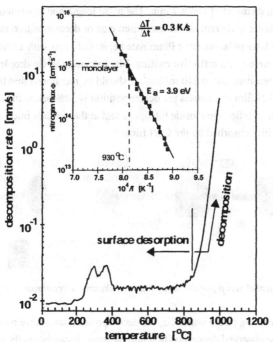

Figure 1. *Thermal induced decomposition of GaN and the corresponding flux of nitrogen from the GaN surface using a heat rate of 0.3 K/s [15].*

LASER INDUCED THERMAL DECOMPOSITION OF GaN

Under normal ambient the thermal stability of GaN is limited by the decomposition of the crystal into nitrogen gas and liquid gallium ($2GaN(s) \rightarrow N_2(g) + 2Ga(l)$). The flux of nitrogen atoms, $\Phi(N)$, leaving the crystal surface increases exponentially above 830°C and can be described by

$$\Phi(N) = 1.2 x 10^{31} \exp(\frac{-3.9eV}{kT}) \frac{1}{cm^2 s}, \tag{1}$$

where kT is the thermal energy (Fig.1) [15]. The corresponding decomposition rate of GaN reaches one monolayer per second at a temperature of 930°C. This result confirms the fact that the decomposition temperature of GaN is much below the melting point and indicates that GaN can be etched by thermal decomposition by methods which enable a localized heating of the sample to temperatures above 830°C.

One way to decompose GaN is to use light with photon energies above the bandgap of GaN, e.g the 355 nm (3.5 eV) third harmonic of a Nd:YAG pulsed laser (Fig.2) [16, 17]. The laser system used in the present work provides a maximum pulse energy of 300 mJ with a Gaussian intensity profile over the beam diameter of about 7 mm. The pulse length and repetition rate is 5 ns and 10 Hz, respectively. In order to determine the maximum etch or decomposition rate of GaN achieved by a single laser pulse we have illuminated epitaxial films with a thickness of 1 μm which were partly covered by a reflective coating. The formation of Ga-droplets was visible for laser output pulse intensities above 310 mJ/cm^2. It should be mentioned that the intensity which is absorbed by the GaN film and causes the decomposition is 30% below the output intensity because of losses due to reflections inside the optics and at the air/GaN interface. The following values are the intensities absorbed by the GaN films.

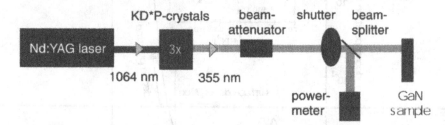

Figure 2. *Experimental setup for the laser induced thermal decomposition of GaN.*

After wet chemical etching of the liquid Ga, the decomposition rates were measured by surface profiling (Fig.3). The observed decomposition rate increases logarithmically with increasing pulse energy above a threshold of about 220 mJ/cm^2 reaching a rate of 100 monolayer per pulse at 290 mJ/cm^2. This corresponds to an optical extinction coefficient a of about 8x10^4 cm^{-1}. For intensities above 320 mJ/cm^2 the increase of etch rate slows down, probably because of the change in optical properties of the GaN during the decomposition of the surface layers. The maximum etch rate per pulse is limited by the output power of our laser system to 90 nm at an absorbed intensity of about 500 mJ/cm^2. Assuming that the liquid Ga can be removed between two laser pulses with a repetition rate of 12 Hz an etch rate of more than 1 μm/s can be achieved well in excess of other etching methods.

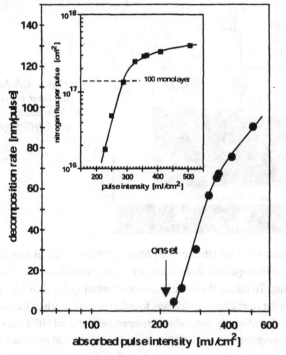

Figure 3. *Decomposition rate and nitrogen flux from the GaN surface versus the absorbed light intensity of single 5 ns laser pulses with a photon energy of 3.5 eV. At intensities of 290 mJ/cm² a decomposition rate of 100 monolayer is measured.*

FREE-STANDING GaN SUBSTRATES

By irradiation of GaN through a transparent substrate, a method of splitting this interface was developed [17]. One application of this method is the fabrication of free-standing GaN substrates starting with sufficiently thick HVPE-films [18]. The laser-induced liftoff of thick GaN layers from the sapphire substrate was performed with the same laser system mentioned above. The pulse energy at the sample surface was adjusted with a polarizing beam attenuator and GaN areas larger than the attainable beam size was achieved by sequentially scanning the pulses across the backside of the sapphire. To protect the film during processing, the film side of the GaN/Al$_2$O$_3$ structure was temporarily bonded to a metal support using a metal solder during the liftoff. Incident pulse intensities of about 400 mJ/cm² were required for the interface splitting at room temperature. Separating areas larger than the attainable beam size was achieved by sequentially scanning the pulses across the backside of the sample, as shown schematically in Fig.4.

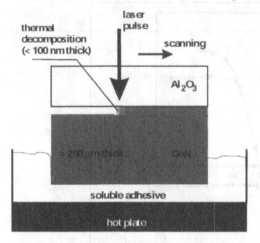

Figure 4. Schematic diagram of the laser process for removal of the sapphire substrate from thick HVPE-grown GaN films.

The bowing of the samples with thicker GaN films conflicted with this procedure, however, because the strain inhomogeneity at the boundary of the released and thus relaxed spots caused extensive fracturing. To reduce the bowing-induced fracturing, we then investigated heating of the sample during the laser liftoff, which was found to reverse much of the strain. The bowing was visibly reduced, and hardly noticeable at temperatures around 600°C. By performing the laser treatment at temperatures above 600°C, extended areas could be released sequentially without sample damage. In this way, free-standing GaN films were separated from 2 inch sapphire wafers, as shown in Fig.5 for a film of 275 μm thickness. On a microscopic scale (10×10 μm^2) a very smooth surface morphology can be observed by atomic force microscopy, with a rms roughness of about 0.4 nm and a dislocation density between 5×10^6 and 2×10^7 cm^2. On a macroscopic scale of several mm the peak-to-peak roughness was determined to be several μm.

Figure 5. Image of a 275 μm thick free-standing GaN film, after removal from the 2 inch sapphire substrate.

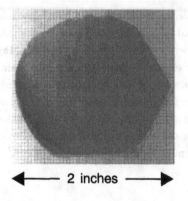

2 inches

An important characteristic of the films after liftoff is that they exhibit little or no bowing, greatly facilitating their application in homoepitaxy or lithography, discussed in the next chapter.

HOMOEPITAXY OF GaN

To prepare smooth surfaces of Ga-face GaN substrates for homoepitaxy, methods like chemical assisted ion beam etching (CAIBE) or reactive ion etching (RIE) were optimized [14]. In this work we were able to improve the surface morphology after mechanical treatment by reactive ion etching (RIE) with BCl_3. For this purpose and after laser-induced delamination of the GaN substrate, the film was mechanically polished with diamond paste until the rms roughness was below 10 nm over areas of several mm^2. Details on the polishing procedure are published elsewhere [19].

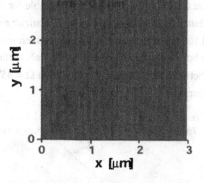

Figure 6. Atomic force microscope image of GaN grown by MOCVD on free standing GaN substrates.

The homoepitaxy was carried out by low-pressure MOCVD using a home-made quartz reactor [19]. Epitaxial films of 2 μm thick GaN were grown on the Ga-face side of the substrate with a growth rate of 600 nm/h at a substrate temperature of 1050°C. The atomic force microscopy (AFM) image shows a very smooth surface morphology of the overgrown GaN (Fig.6). In a cross- section of this image it is possible to identify parallel bilayer steps with a terrace-width of about 200 nm. In addition, based on a series of AFM micrographs we were able to determine the dislocation density to be between 5×10^6 and 2×10^7 cm^{-2}, which is similar to the dislocation density of the original HVPE-GaN substrate. High-resolution x-ray diffraction measurements were performed in order to check the structural quality, strain and mismatch between the GaN substrate and epitaxial layer. 2θ-ω scans of the (0002), (0004) and (0006) reflections were measured with FHWMs down to 20 arcsec. From the high dispersion (0006) diffraction peak and reciprocal space maps of the (205) reflection, the maximum lattice mismatch between substrate and homoeptaxial GaN film was determined to $\Delta a/a < 3 \times 10^{-5}$. To evaluate the optical quality of the GaN homoepitaxial layer, radiative excitonic recombination was measured by photoluminescence spectroscopy (PL) at 4.3 K. The high optical quality of the GaN film is confirmed by the presence of the free exciton lines X_A and X_B and a donor-bound exciton (D^0, $X_A^{n=1}$) with a FWHM of down to 0.5 meV [20]. The optical and structural properties of the free-standing GaN samples fabricated by laser-induced thermal liftoff and overgrown homoepitaxially by MOCVD are very suitable for further applications. Because of the smooth surface and low concentration of free carriers (5×10^{15} cm^{-3}), the samples can be used as substrates for high power, high frequency microwave devices. In addition, because of the low dislocation density, blue laser diodes can be fabricated with long lifetimes, which can also profit from the possibility of producing laser facets by cleaving.

PULSED-LASER INDUCED HOLOGRAPHIC PATTERNING

To realize GaN surface morphologies suitable for antireflection coatings, highly absorbing layers or distributed feedback lasers, striped, rectangular or sine wave patterns with periods between 100 and 500 nm are of interest. To realize such small period patterns the laser beam was split into two beams and superimposed at the GaN surface to achieve a holographic grating [16]. Angles between the interfering beams from $\Omega = 45°$ to $90°$ were used to produce the relevant grating periods.

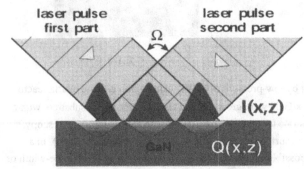

Figure 7. *A grating is produced by splitting the laser beam and bringing the two parts together on the sample surface to make a striped interference pattern. The GaN is decomposed in areas of constructive interference for intensities above 0.2 J/cm².*

For our striped interference patterns with period p, the light intensity in the sample has a profile

$$I(x,z) = I_0 \cos^2(\pi x / p)\exp(-\alpha z), \tag{2}$$

and the absorbed energy density is

$$Q(x,z) = \alpha I(x,z). \tag{3}$$

were α is the absorption coefficient of GaN at the photon energy of the laser pulse. The maximum intensity at the surface, I_o, is reduced from the incident intensity by reflection at the surface, about 20-30% at normal incidence to 420 mJ/cm² for a 100 mJ pulse. Note that this includes a doubling of maximum intensity due to the interference pattern compared to a homogeneous beam profile.

For an absorption coefficient $\alpha = 10^5$ cm^{-1}, the 100 mJ pulse then corresponds to a maximum Q of 40 kJ/cm^3. The heat capacity for GaN has been given in Ref. [21] as

$$C_p(T) = 9.1 + (2.15x10^{-3})T (cal/molK).\qquad(4)$$

Using this relation gives a required heat input of 2.9 kJ/cm^3 to raise the temperature to 900°C, which is 70°C above the decomposition temperature of GaN. This is apparently a small part of the energy delivered by the 100 mJ pulse. Based on vapor pressure measurements, an estimate of decomposition enthalpy at 1300 K of 218.6 kcal/mol N$_2$ was obtained [22]. Using the GaN density of 6.1 g/cm^3 and a molar mass of 83.7 g/mol, this converts to 33 kJ/cm^3 or 4.8 eV per nitrogen atom, which is slightly above the activation energy of the thermal decomposition in e.q. (1). A 210 mJ/cm^2 pulse was just above the threshold for structuring, thus corresponding to a maximum Q near 27.6 kJ/cm^3. If we assign $Q_{max} = \alpha I_0$ to 30 kJ/cm^3 for the 210 mJ/cm^2 pulse and 60 kJ/cm^3 for $I_0 = 420$ mJ/cm^2, this implies an effective absorption coefficient of about 1.4x10^5 cm^{-1}, in good agreement with data from the literature [23].

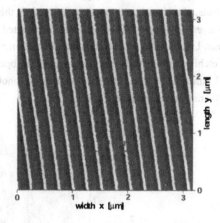

Figure 8. AFM image for a 250 nm period structure in GaN fabricated by laser induced holographic patterning.

Figure 8 shows an atomic force microscope picture of a surface pattern obtained with a single laser pulse (with a peak intensity of 485 mJ/cm^2), using an incidence angle of 45° of the split laser beams. Trenches with a width of about 110 nm and a typical depth of 20 nm were obtained. Even for this nearly diffraction-limited resolution, flat mesa structures, whose sharp edges mark a threshold energy density are produced. The very fine structuring, as in Fig.8, raises the issue of resolution limitation by heat diffusion during the laser pulse. Measurements of the thermal conductivity for room temperature yield $\kappa = 1.3$ W/cmK [24], giving a diffusivity of $D = \kappa/C_p\rho = 0.45$ cm^2/s. Assuming that the thermal conductivity at higher temperatures decreases proportional to T$^{-1.2}$, the heat diffusivity can be estimated to about 0.1 cm^2/s at 1200 K. A cylindrical distribution of temperature with a Gaussian profile of 100 nm radius (corresponding to the absorbed energy profile of a 250 nm grating period) would double this radius in less than 1 ns, and expand to 460 nm during the 5ns pulse duration. If heat diffusivity was this strong, the trenches ought to be broader and flatter. Although the depth of the patterns is apparently reduced by heat diffusion, we conclude that the illuminated areas of GaN reach a temperature during the pulse where the energy input rate is matched by the decomposition rate, so heat is largely removed where it is generated before it can thermally diffuse. This enables the etching of structures with dimensions down to 100 nm by pulsed holographic patterning.

In order to etch deeper structures, we have also investigated the use of multiple pulses. The main obstacle to this approach is the residual gallium, which changes the optical response of the sample for succeeding pulses. Indeed, simply applying 20 pulses at 2 s intervals gave trenches of only three times the depth of a single pulse trench. It is necessary to remove the Ga between pulses, as by exposing the sample to HCl vapor. This method was successful. Exposures to eight pulses spaced 30 s apart produced maximum depth near 400 nm, thus increasing almost linearly with pulse number. A depth of 1 mm was achieved in 2-3 seconds at 10 Hz [25].

If the samples should not be exposed to HCl vapor, e.g. because of the presence of metal contacts, photoresist patterning can be used. In this procedure a smooth GaN film is covered by a thin layer (100 nm) of photoresist and illuminated by the interference patterns of the split laser beams. Like in other photolithography, the illuminated parts of the photoresist are removed by wet etching and the GaN is etched through the open stripes of the photoresist by RIE. By etching for 3 min with BCl_3 a sine shaped surface morphology with a period of 260 nm and an amplitude of 200 nm was fabricated (Fig.9).

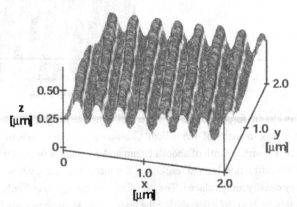

Figure 9. Sinosoidal surface morphology of GaN after patterning using a striped interference pattern to illuminate a thin layer of photoresist and reactive ion etching. Similar results can be obtained by multipulse etching using laser induced thermal decomposition of GaN.

Optimization of multipulse processing will likely determine the attainable etch rates of the decomposition method as well as the ultimate attainable sidewall steepness, within the optical diffraction limit. Besides using interference patterns to rapidly produce one- or two-dimensional arrays of mesas and period structures, more general patterns could be similarly obtained by appropriate masking with protective reflective or ablative material, as well as by direct writing with a focussed beam.

CONCLUSIONS

In conclusion, free-standing high quality GaN substrates as large as 2 inches in diameter can be reproducibly obtained by applying the laser-induced liftoff process of thick GaN HVPE layers. These crack-free bulk-like films are very suitable as substrates for nitride homoepitaxy after careful pretreatment of the Ga-face surface. This was confirmed by the growth of a 2 μm thick GaN film by MOCVD. As observed by AFM and HRXRD measurements, lattice-matched samples were realized with dislocation densities and surface roughness of below 10^7 cm^{-2} and 0.2 nm, respectively. In combination with the good optical and electrical characteristics, these results validate the delaminated HVPE-GaN films as convenient substrates for the fabrication of GaN-based devices. By laser-induced thermal decomposition of GaN using interference patterns, one or two dimensional arrays of mesas and sinusoidal shaped GaN surfaces with periods down to 250 nm were realized. Optimization of single and multipulse processing lead to etch rates of up to 90 nm per pulse and trenches with good sidewall steepness and a depth of 400 nm by using eight pulses. The potential to realize more complicated patterns by using appropriate masking during the laser-induced decomposition or by combining this process with reactive ion etching is demonstrated for surface patterns suitable for antireflection coatings or distributed feedback lasers.

ACHNOWLEDGEMENTS

The authors gratefully acknowledge the collaboration with R. P. Vaudo and V. M. Phanse from ATMI, Inc. in achieving free-standing HVPE-GaN substrates. This work was supported by the Deutsche Forschungsgemeinschaft (Stu 139/5-1 and SFB 348).

REFERENCES

1. H. Teisseyre, P. Perlin, T. Suski, I. Grzegory, S. Porowski, J. Jun, A. Pietraszko, and T.D. Moustakas, J. Appl. Phys. **76**, 2429 (1994).
2. S. Nakamura and G. Fasol, *The Blue Laser Diodes*, Springer, Heidelberg (1997).
3. J.I. Pankove and T.D. Moustakas, *GaN*, Academic Press, New York (1998).

4.   O. Ambacher, J. Phys. D: Appl. Phys. **31**, 2653 (1998).

5.   N.G. Weimann, L.F. Eastman, D. Doppalapudi, H.M. Ng, and T.D. Moustakas, J. Appl. Phys. **83**, 3656 (1998).

6.   S. Nakamura, M. Senoh, S. Nagahama, N. Iwasa, T. Yamada, T. Matsushita, H. Kiyoku, Y. Sugimoto, T. Kozaki, H. Umemoto, M. Sano, and K. Chocho, Jpn. J. Appl. Phys. **36**, L1568 (1997).

7.   F.A. Ponce, D.P. Bour, W. Götz, N.M. Johnson, H.I. Helava, I. Grzegory, J. Jun, and S. Porowski, Appl. Phys. Lett. **68**, 917 (1996).

8.   S. Porowski, I. Grzegory, and J. Jun, *High Pressure Chemical Synthesis*, ed. J. Jurczak and B. Baranowski (Elsevier, Amsterdam, 1989), p. 21.

9.   M. Leszczynski, B. Beaumont, E. Frayssinet, W. Knap, P.Prystawko, T. Suski, I. Grzegory, and S. Porowski, Appl. Phys. Lett. **75**, 1276 (1999).

10.   C. Kirchner, V. Schwegler, F. Eberhard, M. Kamp, K.J. Ebeling, K. Kornitzer, T. Ebner, K. Thonke, R. Sauer, P. Prystawko, M. Leszczyski, I. Grzegory, and S. Porowski, Appl. Phys. Lett. **75**, 1098 (1999).

11.   R.P. Vaudo, V.M. Phanse, X. Wu, Y. Golan and J.S. Speck, 2nd Int. Conf. Nitride Semiconductors, Tokushima, 1997.

12.   Y. Golan, X.H. Wu, J.S. Speck, R.P. Vaudo, and V.M. Phanse, Appl. Phys. Lett. **73**, 3090 (1998).

13.   S. Nakamura, et al.,Appl. Phys. Lett. **72**, 2014 (1998); Appl. Phys. Lett. **73**, 832 (1998).

14.   S.J. Pearton, J.C. Zolper, R.J. Shul, and F. Ren, J. Appl. Phys. **86**, 1 (1999).

15.   O. Ambacher, M.S. Brandt, R. Dimitrov, T. Metzger, M. Stutzmann, R.A. Fischer, A. Miehr, A. Bergmaier and G. Dollinger, J. Vac. Sci. Technol. B**14**, 3532 (1996).

16.   M.K. Kelly, O. Ambacher, B. Dahlheimer, G. Groos, R. Dimitrov, H. Angerer, and M. Stutzmann, Appl. Phys. Lett. **69**, 1749 (1996).

17.   M.K. Kelly, O. Ambacher, R. Dimitrov, R. Handschuh, and M. Stutzmann, phys. stat. sol. (a) **159**, R3 (1997).

18.   M.K. Kelly, R.P. Vaudo, V.M. Phanse, L. Görgens, O. Ambacher, and M. Stutzmann, Jpn. J. Appl. Phys. **38**, L217 (1999).

19.   C.R. Miskys, M.K. Kelly, O. Ambacher, and M. Stutzmann, phys. stat. sol. (a) **176**, 443 (1999).

20.   B.J. Skromme, J. Jayapalan, R.P. Vaudo, and V.M. Phanse, Appl. Phys. Lett. **74**, 2354 (1999).

21.   I. Barin, O. Knacke, and O. Kubaschewski, *Thermochemical Properties of Inorganic Substances* (Springer, Berlin, 1977).

22.   Z.A. Munir and A.W. Searcy, J. Chem. Phys. **42**, 4223 (1965).

23.   D. Brunner, H. Angerer, E. Bustarret, F. Freudenberg, R. Höpler, R. Dimtrov, O. Ambacher, and M. Stutzmann, J. Appl. Phys. **82**, 5090 (1997).

24.   E.K. Sichel and J.I. Pankove, J. Phys. Chem. Solids **38**, 330 (1977).

25.   M.K. Kelly, O. Ambacher, R. Dimitrov, H. Angerer, R. Handschuh, and M. Stutzmann, Mat. Res. Soc. Symp. Proc. **482**, 973 (1998).

Laser-Driven
Nanoparticle Formation

Mat. Res. Soc. Symp. Vol. 617 © 2000 Materials Research Society

Deposition of Nanotubes and Nanotube Composites using Matrix-Assisted Pulsed Laser Evaporation

P. K. Wu[1], J. Fitz-Gerald[2], A. Pique[2], D.B. Chrisey[2], and R.A. McGill[2]

[1] Southern Oregon University, Ashland, OR.
[2] Naval Research Laboratory, Code 6370, Washington D.C. 20375

Using the Matrix-Assisted Pulsed Laser Evaporation (MAPLE) process developed at the Naval Research Laboratory, carbon nanotubes and carbon nanotube composite thin films have been successfully fabricated. This process involves dissolving or suspending the film material in a volatile solvent, freezing the mixture to create a solid target, and using a low fluence pulsed laser to evaporate the target for deposition inside a vacuum system. The collective action of the evaporating solvent desorbs the polymer/nanotube composite from the target. The volatile solvent is then pumped away leaving the film material on the substrate. By using this technique single-wall-nanotubes (SWN) have been transferred from the target to the substrate. The SWN sustain no observable damage during the deposition process. Using SWN in combination with polymers as the target material, SWN/polystyrene and SWN/polyethylene glycol composite films were made. These films can be deposited on a variety of substrates, e.g., Si, glass, plastic, and metal, using the same target and deposition conditions. SEM micrographs show that the SWN were uniformly distributed in the film. Using a simple contact mask, SWN composite films 20 um diameter patterns can be produced.

I. Introduction

Nanotubes were first identified by Sumjo Iijima.[1] Since their discovery, new fabrication and purification techniques have enhanced the production of single-wall-nanotubes (SWN).[2-4] SWN with lengths up to a few hundred μm can be produced. SWN are interesting materials in many ways. They can be either metallic or semiconducting, depending on its chiral structure and diameter.[5,6] SWN not only have exceptionally high modulus, $> 1 \times 10^{12}$ Pa,[7,8] but are also flexible and strong.[9] These remarkable properties make SWN a subject of intense materials research.

The present work illustrates how SWN/polymer composite thin film can be fabricated using Matrix-Assisted Pulsed-Laser Evaporation (MAPLE).[10-12] We will demonstrate that MAPLE is suitable for fabricating a variety of SWN/polymer composites. This is technologically important because SWN/polymer composite is one way to fabricate materials with custom properties for specific applications. For example, the mechanical toughness for a polymer can be improved by forming a polymer composite with SWN. Because of the stability and inertness of SWN, SWN/polymer is potentially useful as biomaterials in applications such as requiring special structure or in novel drug delivery. Furthermore, forming a composite with SWN can alter the dielectric constant of a polymer; thus improving the polymer's electrical characteristics for applications such as interconnects. In short, the ability to make polymeric composites with SWN permits the alteration of materials properties, i.e., mechanical and electrical. This ability broadens the possible applications for a material as well as the choice of materials for a particular application. The integrity and inertness of SWN make them ideal for most applications, from integrated circuits to bio-applications.

II. Experimental
II. a. Matrix-Assisted Pulsed-Laser Evaporation (MAPLE).

MAPLE is conducted in a conventional vacuum chamber equipped with a window of high transmitance for the wavelength of the pulsed laser to be used. The target is a frozen matrix consisting of a volatile solvent, e.g., water, methanol, chloroform, etc., and a low concentration of the film materials, in this case SNW and a polymer which is dissolved in the solvent. The criteria for choosing the volatile solvent are high absorption of the laser light; high solubility for the polymer of choice; and is volatile at room temperature which allows it to be pumped away quickly. The substrate is placed directly in front of the target. Deposition proceeds by illuminating the target with laser pulses. The laser pulse initiates a photo-thermal process, vaporizing the frozen solvent, and releasing the solute and SWN into the chamber. The momentum, resulting from the vaporization process, carries the solvent, solute, and SWN to the substrate. Because the solvent has a high vapor pressure at room temperature, it will rapidly be removed by the pumping system. The polymer and SWN will then form a continuous and dry film. The key to the MAPLE process is that the laser interaction occurs primarily with the solvent molecules, the polymer and SWN should be undamaged and remain intact.[10-12]

II b. Deposition Conditions

In the current work, films were deposited using λ = 248 or 193 nm. The fluence at the target is 0.15 - 0.25 J/cm^2, with a pulse repetition rate from 1 to 10 Hz. Substrate temperature is 300 K and a LN2 assembly cools the rotating target. The base pressure of the vacuum system is 10^{-5} Torr and the system is backfilled to 50-200 mTorr of Ar during deposition. The substrate-to-target distance is 10 cm. Si wafers, glass slides, Cu sheets, and Kapton sheets were used as substrates. Patterning is done using a contact mask attached directly to the substrate.

The SWN were supplied by a special Office of Naval Research (ONR) program with Rice University. The as-received SWN are suspended in toluene with a trace amount of NaOH and some particulate of soot or graphite. In the current work, the toluene is first removed by evaporation or pumping and replaced by chloroform. To make the MAPLE target, 1.1 gm of polymer, either polystyrene (PS) or polyethylene glycol (PEG), is mixed with 25 gm of solution. Because the ratio of SWN to chloroform cannot be determined, the exact ratio of SWN/polymer in the target is not known.

III. Results and Discussion

A drop of the as-received SWN solution is allowed to dry on a piece of Si forming a SWN film. A scanning electron micrograph (SEM) from this sample is shown in Figure 1a. The SWN can be easily identified in this figure. The SWN are smooth and without a preferred orientation. Fine particulates, possibly NaOH, graphite, or carbon particles, can be observed on the SWN. Some of the SWN are twisted and bundled together forming larger diameter "ropes".

Figure 1b shows an SWN film fabricated using MAPLE and with the as-received SWN with chloroform in the target. Again the SWN can be easily identified. The concentration of the SWN in this film is much lower than that shown in Figure 1a. Because the solvent for the MAPLE target is chloroform, which is volatile at room temperature, we believe the matrix in this film is NaOH with some graphite and/or carbon particles. A comparison of the SWN in Figure 1a and b shows that the thickness of the SWN is preserved through the MAPLE process.

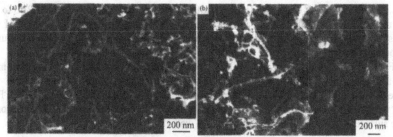

Figure 1.Surface of a SWN film on Si, formed by a) drying as-received SWN on Si and   b) MAPLE with chloroform as the target solvent.

The SWN shown in Figure 1b do not show the same twisting and bundling as that in Figure 1a. This result indicates that, under the current deposition conditions, the MAPLE process mainly involves individual SWN. Individual SWN arrive at the surface and solidify before mixing occurs.

A scratch is made on the film in figure 1b, and SEM from this crack are shown in Figure 2. These images show that the SWN are imbedded in the matrix materials and not just on the surface. The fragmented film is connected by SWN, see Figure 3b. This illustrates that the film is truly a composite material that should exhibit mechanical benefits associated with composites.

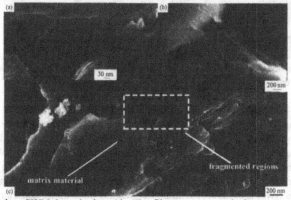

Figure 2 a, b, and c. SWN deposited on Si. The film was purposely fragmented to show SWN nanotubes bridging fragmented matrix sections.

The SWN connecting the two cracked pieces are on the order of 0.5 μm indicating that the SWN should be much longer than that, 1.0 μm. Longer SWN can be found on the upper right corner of Figure 2b. These SEM show that SWN can be transferred without breakage and more generally, that MAPLE can transfer very large molecular species, $> 1 \times 10^6$ amu, intact. To the best of our knowledge, MAPLE is the only deposition process that combines all the advantages of a physical vapor deposition technique, is "gentle" to polymers, [10-12] and can make films with materials of the size and extent as the SWN.

Figure 3a – 3d show SEM from the surface of a composite SWN/PEG film fabricated by

MAPLE. Again, SWN can easily be identified further demonstrating the ability of MAPLE to transfer macro structures. The twisting and bundling of SWN are not observed in all the SEM from this sample. Thus the MAPLE deposition proceeds as a molecule-by-molecule process under our deposition conditions.

Figure 3a – 3c clearly show the SWN embedded in the PEG matrix. The film shown in Figure 3a and 3b appears continuous, homogenous, and without void. This may indicate that when the SWN and PEG arrive at the sample surface, the PEG is still in the liquid state. The PEG flows to wet and fill voids in the film before solidification. The film is free of SWN protrusions from the surface indicating the PEG wets the SWN well.

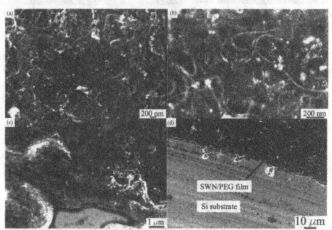

Figure 3a – 3d. Surface of a SWN/PEG film on Si made by MAPLE with chloroform as the target solvent.

Figures 4a-4c show SEM from the surface of a SWN/PS film fabricated by MAPLE. Again, the SWN can be easily identified showing good target-to-sample transfer characteristics. The structure of the SWN/PS film shown in Figure 4 indicates that the growth mode of SWN/PS is different from the previously shown films, unlike the other films in which the SWN intermixed with the matrix. The SWN in this film are not imbedded in a PS matrix. Instead a three-dimensional network of PS coated SWN is formed on top of a film composed mainly of PS.

The fact that SWN is capable of forming freestanding three-dimensional structures means that they are structurally reinforced by the PS when they arrive at the surface. Thus the PS either never evaporates completely from the SWN at the target or the SWN act as nucleation and growth sites for the PS during the transfer process. When the SWN arrive on the sample surface, they are coated with PS. The SWN get stuck on the sample surface or other SWN already present on the surface, largely preserving their orientation and shape they acquired. PS transferred subsequently further secures these structures. PS that is transferred subsequently and that does not encounter SWN will be deposited on the substrate forming the observed PS film underneath.

SWN, which are much thicker than a single strand SWN, can be identified, see Figure 4b and 4c. However it is not clear at this point that these are bundled SWN or single strand SWN with a thick coating of PS. Figures 4b and 4c show high-resolution images of these features. The "icicle" structure closely resembles an ice formation on branches during periods of high humidity. This is a clear indication of growth from the vapor phase. The observed large diameter of some of the SWN can thus be a result of PS growth rather than bundling of SWN.

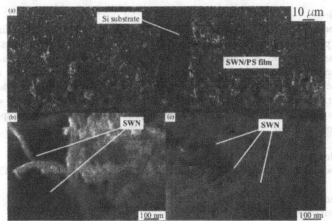

Figure 4a - c. Surface structures of a SWN/PS film on Si made by MAPLE with chloroform as the target solvent.

The difference in morphology between the SWN/PS and SWN/PEG can be chemical in nature, i.e., the functional groups on the polymer may have specific reactions with the SWN that cause the observed differences. The differences may also result from parameters such as molecular weight and molecular concentration. Note that our target concentration is weight concentration and not molecular concentration.

Circular SWN/PS dots, ranging from 25-200 μm in diameter, have been successfully fabricated using a contact mask. SEM of these structures are shown in Figure 5a and 5b.

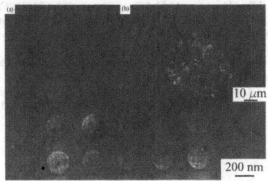

Figure 5a and b. SWN/PS dots ranging from 25-200 μm in diameter deposited on Si substrates.

Considering that the patterning is done by taping a metal mask on the substrate, where gaps between the mask and the substrate should be expected, the definition of the edges is of high quality. In addition to the films discussed above, we have successfully deposited SWN/polymer thin films on glass slides, Cu, and Kapton. These films show characteristics similar to the films discussed here.

IV. Conclusion

Our preliminary results have validated the ability of the MAPLE technique for fabricating SWN, SWN/polymer composites, and patterned SWN/polymer thin films on various substrate materials. Parameters such as polymer molecular weight may have significant influence on the growth rate and morphology of the film, which will require further exploration. We are in the process of measuring the mechanical properties, i.e., toughness and adhesion, and electrical properties, i.e., dielectric constant and resistivity, of these films. These parameters will help identify the potential applications of these materials systems and consequently the specific areas, i.e., fabrication method or composition, which necessitate refinement.

V. Acknowledgement

The authors would like to thank ONR for sponsoring this research and the ONR SWN program.

VI. Reference:

1. S. Iijima, Nature **354,** 56 (1991).
2. T.W. Ebbesen and P.M. Ajayan, Nature **350**, 220 (1992).
3. T.W. Ebbesen, Annu. Rev. Mater. Sci., **24**, 235 (1994).
4. A. Thess, R. Lee, P. Nikolaev, H. Dai, P. Petit, J. Robert, C. Xu, Y.H. Lee, S.G. Kim, D.T. Colbert, G. Scuseria, D. Tománek, J.E. Fischer, and R.E. Smalley, Science, **273**, 483 (1996).
5. R. Saito, M. S Fujita, G. Dresselhaus, and M.S. Dresselhaus, Appl. Phys. Lett., **60**, 2204 (1992).
6. T.W. Ebbesen, H.J. Lezec, H. Hiura, J.W. Bennet, H.F. Ghaemi, and T. Thio, Nature, **382**, 54 (1996).
7. F. Banhart and P.M. Ajayan, Nature, **382**, 433 (1996).
8. M.M.J. Treacy, J.W. Ebbesen, and J.M. Gibson, Nature, **381**, 678 (1996).
9. S. Iijima, C.J. Brabec, A. Maiti, and J. Bernholc, Nature, **104**, 2089 (1996).
10. R.A. McGill, D.B. Chrisey, A. Piqué, and T.E. Mlsna, Mat. Res. Soc. Symp. Proc., **526**, 375 (1998).
11. A. Piqué, D.B. Chrisey, B.J. Spargo, M.A. Bucaro, R.W. Vachet, J.H. Callahan, R.A. McGill, D.Leonhardt, and T.E. Mlsna, Mat. Res. Soc. Symp. Proc., **526**, 375 (1998).
12. A. Piqué, R.A. McGill, D.B. Chrisey, D. Leonhardt, T.E. Mlsna, B.J. Spargo, J.H. Callahan, R.W. Vachet, R. Chang, and M.A. Bucaro, Thin Solid Films, **355-356**, 536 (2000).

Mat. Res. Soc. Symp. Vol. 617 © 2000 Materials Research Society

Novel Nanocrystalline Materials by Pulsed Laser Deposition

J. Narayan, A.K. Sharma, A. Kvit, D. Kumar, and J.F.Muth[1]
NSF Center for Advanced Materials and Smart Structures, Department of
Materials Science and Engineering, North Carolina State University, Raleigh, NC
27695-7916.
[1]Department of Electrical and Computer Engineering, North Carolina State
University, Raleigh, NC 27695.

ABSTRACT

We have developed a novel method based upon pulsed laser deposition to produce nanocrystalline metal, semiconductor and magnetic material thin films and composites. The size of nanocrystals was controlled by interfacial energy, number of monolayers and substrate temperature. By incorporating a few monolayers of W during PLD, the grain size of copper nanocrystals was reduced from 160nm (Cu on Si (100)) to 4nm for a multilayer (Cu/W/Cu/W/Si (100)) thin film. The hardness increased with decreasing grain size up to a certain value (7nm in the case of copper) and then decreased below this value. While the former is consistent with Hall-Petch model, the latter involves a new model based upon grain boundary sliding.

We have used the same PLD approach to form nanocrystalline metal (Ni, Co, Fe embedded in α-Al$_2$O$_3$ and MgO) and semiconductor (Si, Ge, ZnO, GaN embedded in AlN and α-Al$_2$O$_3$) thin films. These nanocrystalline composites exhibit novel magnetic properties and novel optoelectronic properties with quantum confinement of electrons, holes and excitons in semiconductors. We review advanced PLD processing, detailed characterization, structure-property correlations and potential applications of these materials.

INTRODUCTION

Nanocrystalline (nc) materials with grain size 1-100nm range are found to exhibit improved hardness, increased ductility and toughness, superior magnetic and optoelectronic properties [1-5]. However, the properties of nc materials in general and mechanical properties in particular have been found to be controversial due to difficulties in compaction and unavailability of "artifact-free" nc samples [6].

In the present investigation, we have used a modified pulsed laser deposition method to control three-dimensional nucleation or island growth. The size of nanocrystals is determined by interfacial energy, number of monolayers and substrate temperature. We review processing, characterization and structure-property correlations of novel metallic, magnetic and semiconductor nanocrystalline materials.

1) Nanocrystalline Metallic Materials

The basic idea behind processing of nanocrystalline materials is demonstrated in Fig. 1 which shows formation of three-dimensional Ge islands on Si(100) substrate at 400°C. This growth mode is induced when the average of film and substrate interaction energies $(W_{AA} + W_{BB})/2$ exceeds the interaction energy between the film and the substrate (W_{AB}).

In the case of epitaxial Ge growth on Si(100) substrate, there is a transition from two-dimensional layer-by-layer to three-dimensional island growth at a substrate temperature of 375 C. The high-resolution electron micrograph clearly shows (111) planes and facetting in the Ge nanocrystals. It is interesting to note that the size of these islands is quite uniform as the system can lower its energy. The grain size of Cu nanocrystals was controlled by introducing a few interposing monolayers of W which is insoluble in Cu and has a high surface energy with little or no interaction with the film [7, 8]. These characteristics of W have been used to satisfy the energy criterion and induce three-dimensional island growth. The size of the island hence the nanocrystal is controlled primarily by the amount of copper and the substrate temperature. The average grain size of Cu on Si(100) was determined to be 160 nm. By introducing an interposing layer of W between the film and the Si (100) substrate, the grain size was reduced to 80 nm at 10 Hz of laser repetition rate. When the repetition was increased to 15 Hz, the grain size decreased further to 70 nm.

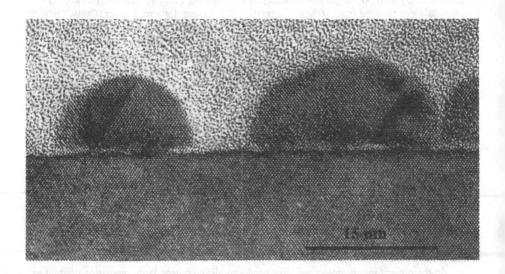

Fig. 1: *High resolution cross-section micrograph in <110> direction showing the Ge islands (nanocrystals) grown at 400°C on Si (100) substrate.*

We achieved a significant reduction in grain size by introducing a multilayer of W/Cu where we controlled the grain size of Cu by the amount of copper deposited. The details of multilayer deposition conditions are summarized in Table I. By choosing proper deposition conditions, grain size was reduced to 4 nm. The cross-section micrograph also demonstrates a high quality of the film without any porosity or voids.

Table I: Details of Deposition Conditions for Cu/W Multilayer specimens

Sample no:	Growth conditions	Grain Size (nm)
Cu/W#6	E= 250 mJ Cu=4min, W= 45sec 10 Hz	20
Cu/W#7	E=261 mJ Cu=3 min, W=20 sec 10 Hz	11
Cu/W#8	E= 260 mJ Cu = 2.5 min W=10 sec 10Hz	7
Cu/W#9	E=260mJ Cu=1.5 min W=10 sec 10Hz	4

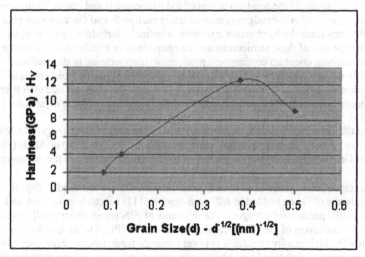

Fig. 2: *Hardness (H) as a function of $d^{-1/2}$ (Hall-Petch relationship), showing some softening with grain size below a critical value.*

The nanohardness measurements as a function of depth were made on all the samples, using the nanoindentation technique. The hardness increased with decreasing grain size attaining the highest value of 12.5 GPa for 7 nm specimen (as in Fig.2). It is interesting to note that the hardness decreased as the grain size decreased to 4 nm. The increase in hardness (H) with the decrease in grain size (d) can be rationalized by Hall-Petch model where H scales with $d^{-1/2}$ [7-9].

However, to explain the hardness decrease with grain size, we have developed a new model based upon grain boundary deformation or sliding [7-9]. According to this model strain rate $\dot{\varepsilon}$ is given by

$$\varepsilon = (\alpha_b A \lambda \eta_b v_b / d). \, Exp(-\Delta G_b / RT). \, Sinh(\tau \Omega / RT) \quad (1)$$

where α_b is a grain boundary constant, η_b number density of atoms in the grain boundary, v_b Debye frequency, ΔG_b activation barrier for jumps, Ω activation value, τ stress, A area swept and λ jump distance. This model predicts decrease in hardness with grain size due to grain boundary sliding. The model has been discussed in detail in our recent papers and a good agreement between the model and experimental results obtained [7-8].

2) Nanocrystalline Semiconductors

Electrons and holes can be confined in semiconductor nanocrystals by surrounding them with higher band-gap materials [10-11]. Using this basic concept of electron-in-a-box, it is possible to increase (blueshift) the bandgap of embedded nanocrystals and obtain luminescence with varying frequency. Indirect bandgap semiconductors such as Si and Ge have very poor luminescent efficiency since the band center transition is optically forbidden. However, by making the crystallite size of these semiconductors comparable to or smaller than the exciton Bohr radius, the resulting quantum confinement produces a sharp increase in the oscillator strength and shifts the luminescence to higher energies. Since the excitonic Bohr radius of bulk Ge [12] is considerably larger (24.3 nm) than that of Si (4.9 nm), the quantum size effects are expected to be more prominent in Ge nanocrystals even for larger sizes of these crystallites [12-14].

By using the same PLD method, we can induce island growth of Ge (bandgap 0.67eV) and ZnO (bandgap ~3.4eV) on wider bandgap materials such as AlN (~6.2 eV) or Al_2O_3 (~9.9 eV). The size of Ge island (nanocrystal) is controlled by the amount of Ge and the substrate temperature.

Pulsed excimer laser system (λ~248 nm, pulse duration=25 ns) was used to deposit up to ten quantum dot layers of Ge embedded in AlN on p-type Si (111) substrate in vacuum and also as a function of partial pressure of nitrogen. The thickness of AlN layers was typically ~40 nm followed by 5 – 6 monolayers of Ge. The pulsed laser deposition (PLD) technique has the advantage of growing high quality structures at lower substrate temperatures. Moreover, we can minimize the possibility of defect induced photoluminescence and contributions from interfacial GeO_x layers more accurately. This allows us to investigate quantum confinement effects in these nanostructures embedded in a high dielectric matrix.

The characterization of these structures was performed using high resolution transmission electron microscopy (HRTEM), photoluminescence (PL) and Raman spectroscopy. The growth conditions for samples grown at different substrate temperatures were found to be quite critical as shown by TEM results. It was observed that these nanodots exhibited crystalline structure when grown at 300 – 500°C. The size of islands increased as a function of deposition temperature and decreased with the thickness of Ge layer. Much lower substrate temperatures such as room temperature resulted in poorly crystalline or amorphous structures. In the high

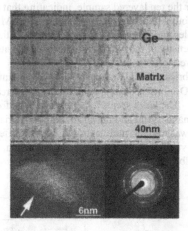

Fig. 3: *Ge nanocrystals embedded in amorphous alumina a-Al₂O₃ and corresponding electron diffraction. The arrow in the high-resolution micrograph of the nanocrystal indicates the growth direction.*

resolution image (Fig. 3) we observe lattice fringes corresponding to {111} planes of Ge with the diamond cubic structure (0.326 nm). The Ge nanodots are pyramidal shaped with height~15nm and base dimensions ~20nm. The HRTEM images reveal that these islands are slightly strained but dislocation free. The Raman spectrum (Fig. 4) reveals a peak downward shifted, upto 295/cm, of the Ge-Ge vibrational mode caused by quantum confinement of the phonons in the Ge-dots.

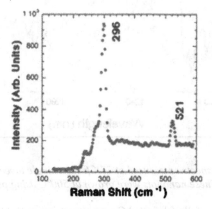

Fig. 4: *Raman spectra from a single layer of Ge nanodots embedded in AlN matrix.*

The PL (Fig. 5) of the Ge dots was blue shifted by ~0.266 eV from the bulk Ge value of 0.73 eV at 77 K, resulting in a distinct peak at ~1.0 eV. The FWHM of the peak was 13 meV, for the single layered and 8 meV for the ten layered sample, indicating that the Ge nanodots are fairly uniform in size, which was found to be consistent with our HRTEM results. The quantum confinement of electrons, holes and excitons [14] was found to be the primary mechanism for the photoluminescence. The intensity of the photoluminescence indicated that the radiative recombination process was very efficient.

Recently, we have fabricated ZnO (bandgap ~3.45 eV) quantum confined nanocrystals embedded in AlN and α-Al₂O₃. In this case, we can blueshift (increase the bandgap) as well as redshift (lower the bandgap). The latter seems to be related to the formation of secondary phases at the interface between the nanocrystal and the surrounding materials. However, the most exciting aspect of this work was to build multilayer stacks with different ZnO sizes which can cover the entire range of the visible spectrum. Using this approach, we were able to produce a high intensity white light bulb.

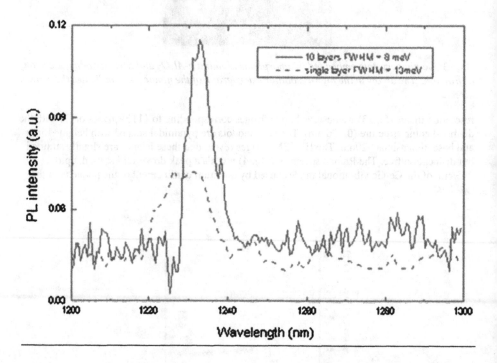

Fig. 5: *Photoluminescence spectra measured at 77K from a single layer and ten layered Ge embedded in AlN matrix. These nanodots were grown at 500°C using pulsed laser deposition.*

Thus, we have successfully fabricated Ge nanostructures embedded in AlN matrix with fairly uniform size distribution. Quantum confinement effect was confirmed by the distinct blue shifted photoluminescence peak observed at ~1.0 eV at 77 K. The sharp FWHM of the PL peak consisting of ten distinct quantum dot layers suggests a uniform size distribution of the nanodots. The PLD method and the use of AlN matrix demonstrate the ability to build multilayers of nanodots in a controlled manner and place them in an oxygen free environment at low substrate temperatures.

3) Nanocrystalline Magnetic Materials

Nanocrystalline magnetic materials in the size range 10-20nm can acquire the properties of single domains even in zero magnetic field [15]. The magnetic axis of a single-domain nanocrystallite is determined by the magnetic anisotropy energy (CV), where C is the anisotropy energy per unit volume and V is the volume of the particle. The parallel and anti-parallel orientations along the magnetic axis are degenerate energetically with an energy barrier CV. By controlling the V or size of nanocrystallites, it is possible to tune the properties of nanocrystalline composites from superparamagnetic to ferromagnetic behavior. In other words, we can control the blocking temperature (T_B) by manipulating the size of the magnetic nanocrystallites. Superparamagetic properties of single-domains are characterized by the relaxation time (τ):

$$\tau = \tau_0 \exp\left(\frac{CV}{kT}\right) \tag{2}$$

From this, the blocking temperature (T_B) above which the magnetic anisotropy energy barrier is overcome by thermal energy kT, is given by

$$T_B = CV/k \ \ln(\tau/\tau_0) \tag{3}$$

Where 'τ' is a function of instrument sensitivity. Above T_B, the nanocrystalline materials exhibit superparamagnetic behavior. Ferromagnetic properties below T_B are characterized by spontaneous magnetization, coercivity and remanence.

Magnetic susceptibility (χ) of superparamagnetic materials above T_B is given by

$$\chi = PV \ \frac{M_s^2}{3kT} \tag{4}$$

Where 'P' is the volume fraction of magnetic nanocrystalline material. The spontaneous magnetization $M_s(T)$ as a function of temperature follows

$$M_s(T) = M_0 (1 - BT^{3/2}) \tag{5}$$

Where B is a constant, and M_0 is the volume at T = 0.

The coercivity $H_c(T)$ for T < T_B is given by

$$H_c(T) = H_c(0) \left(1 - \sqrt{T/T_B}\right) \tag{6}$$

In nanocrystalline materials, a significant enhancement in coercivity can occur at a particular volume fraction P of the nanocrystalline material.

Fig. 6 shows Ni nanoparticles embedded in α-Al$_2$O$_3$. It is interesting to note that the size of these quasispherical nanoparticles is fairly uniform. The size is primarily determined by the substrate temperature and the amount of the deposited nickel. The magnetic measurements in Fig. 7 clearly show ferromagnetic behavior (coercivity and remnant magnetization) below T$_B$, and paramagnetic behavior (zero coercivity and magnetization) above T$_B$. The T$_B$ from the M-H curves is estimated to be 30K.

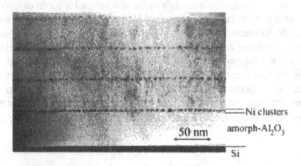

Ni-clusters (#2) in amorphous Al$_2$O$_3$

Fig. 6: *Cross-sections TEM micrograph showing Ni nanocrystals in amorphous aluminum oxide.*

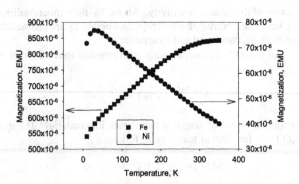

Fig. 7: *Magnetization (M) as a function of field (H) at different temperatures.*

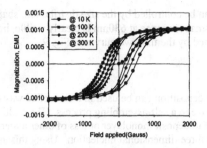

Fig. 8: *Zero-field-cooled and field-cooled M vs. Temperature (T).*

Fig. 8 shows M-H plots where comparison between zero-field-cooled (ZFC) and field-cooled (FC) behavior is clearly delineated. The point of separation between the ZFC and FC characteristics corresponds to T_B, in agreement with M vs H measurements of Fig. 7. Using iron nanoparticles, it was possible to raise T_B in the range of 300K. For many magnetic applications such as high density recording, the amplitude and width are proportional to $(M_r H_c)^{1/2}$ and $(1/H_c)^{1/2}$, respectively, where H_c is the coercivity and M_r is the remnant magnetization. Thus, larger H_c and M_r result in better recording characteristics.

4) Nanocrystalline Optical Materials

The nanocrystalline metallic materials embedded in a wide-band-gap semiconductor or ceramic such as MgO, α-Al_2O_3, SiO_2 can modify their optical properties in a interesting way [16]. The Au and Ag particles embedded in α-Al_2O_3 matrix exhibit strong surface plasmon resonance in the visible range as shown in Fig. 9.

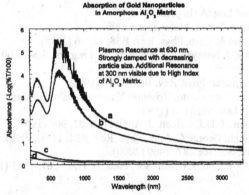

Fig. 9: *Optical absorption as a function of wavelength showing plasma resonance associated with Au nanocrystals. Curves c,d correspond to small Au crystals.*

The absorption characteristics can be controlled by the size distribution, shape and volume fraction of the particles. The plasma resonance at 630nm is clearly visible, but it is damped as the particle size decreases, as exhibited by the first two curves in Fig. 9.

CONCLUSIONS

In summary, pulsed laser deposition can be used in a controlled way to obtain three-dimensional islands of uniform size in a self-assembling process. The multi-target configurations of PLD lend itself to deposit monolayer or less of other material as a surfactant to control the interfacial energy for three-dimensional nucleation. Using this approach, we have fabricated nanocrystalline metals (Cu, Zn) and their composites, semiconductor nanocrystals embedded in ceramics or wide bandgap semiconductors, and metal-ceramic composites for improved and novel optical and magnetic properties.

ACKNOWLEDGEMENTS

We gratefully acknowledge National Science Foundation for the support of this research.

REFERENCES

1. J. Narayan, Y. Chen, and R.M. Moon, Phys. Rev. Lett. **46**, 1491 (1981).
2. J. Narayan and Y. Chen, Phil. Mag. **A49**, 475 (1984; U.S. Patent #4,376,455 (March 15, 1983).
3. J. Narayan, Y. Chen, and K.L. Tsang, Phil. Mag. A55, 807 (1987).
4. C. Suryanarayana Int. Mater. Review **40**, 41 (1995).
5. H. Gleiter, Prog. Mater. Sci., **33**, 223 (1989).
6. C. Koch in Nanostructure Science and Technology ed. By R.W. Siegel, E. Hu and M.C. Rocco (1999) p.93.
7. J. Narayan, Mat. Sc. and Eng. A (in press).
8. H. Conrad and J. Narayan, Script. Mater. (Jan. 2000).
9. H. Van Swygenhoven and A. Caro, Phys. Rev. **B58**, 11246 (1998).
10. K. Hassan, A.K. Sharma, J. Narayan, J.F. Muth, C.W. Teng and R.M. Kolbas, Appl. Phys. Lett. **75**, 1222 (1999).
11. B. Delley and E.F. Steigmeier, Phys. Rev. **B47**, 1397 (1993).
12. Yoshihito Maeda, Nobuo Tsukamoto, Yoshiaki Yazawa, Yoshihiko Kanemitsu and Yasuaki Masumoto, Appl. Phys. Lett. **59**, 3168 (1991).
13. A.G. Cullis, L.T. Canham, P.D.J. Calcott, J. Appl. Phys. **82**, 909 (1997).
14. V. Ranjan, Vijay A. Singh, George C. John, Phys. Rev. **B58**, 1158 (1998).
15. M.J. MacLachlan, et al., Science **287**, 1460 (2000).
15. J.A. Creighton and D.G. Eadon, J. Chem. Soc. Faraday Trans., **87**, 3881 (1991).

Mat. Res. Soc. Symp. Vol. 617 © 2000 Materials Research Society

Optical and Structural Characteristics of Gold Nanocrystallites Embedded in a Dielectric Matrix

A.K. Sharma, J.F. Muth[1], A. Kvit, J. Narayan and R.M. Kolbas[1]
Department of Materials Science and Engineering, North Carolina State University,
Raleigh, NC 27695-7916.
[1]Department of Electrical and Computer Engineering, North Carolina State University,
Raleigh, NC 27695.

ABSTRACT

We have fabricated Gold (Au) crystallites in amorphous alumina matrix by pulsed laser deposition. The characterization of these multilayer sequences was performed by high resolution transmission electron microscopy (HRTEM) and optical transmission measurements. TEM studies revealed the morphology and the microstructure of these composites. The optical transmission spectra showed characteristic surface plasma resonance of Au particles confined by the host dielectric matrix. These resonances fall in the visible spectral range. The importance of pulsed laser deposition in fabricating these composite films is discussed.

INTRODUCTION

The study of colloidal metal particles has a long and distinguished history due to the strong optical resonances in the visible portion of the spectrum permitting the design of colored glasses for stained glass in cathedrals and artware [1]. The Mie theory of absorption, which explains much of the fundamental characteristics of colloidal particles, was published in 1908 [2]. Since then there has been substantial work in many fields concerning the optical characteristics of small particles and aerosols [3]. Many methods [4] have been developed to produce these particles including: precipitation in solution, sol-gel, ion implantation, precipitation of metal clusters via annealing, chemical synthesis, co-sputtering, electron–beam lithography and low energy cluster-beam [5]. Advances in computer technology have also made available various computer codes, which may be used to calculate the resonances of non-spherical particles [3].

The optical response of metallic nanoparticles has a direct correspondence to the particles electronic structure and is strongly dependent on the particle size and shape, which create what are termed Morphological Dependent Resonances (MDRs). The optical absorption bands of noble metals such as gold, and silver fall in the ultraviolet-visible spectral range. The strong absorption is the manifestation of the excitation of their characteristic surface plasmon resonance or interband transitions. The spectral position of surface plasma resonance of the metallic particles can be varied by embedding them in appropriate dielectric matrices. Controlling the size, shape and the composition of the surrounding matrix provides us a way to tailor the optical properties of the metal-dielectric composite as a function of the size of the metal particles and their volume fractions. Thus to better understand the behavior of metallic nanoparticles methods of creation which permit the size and shape and the encapsulating environment to be controlled are strongly preferred. In conjunction with refinements of the deposition method it is important to be able to directly measure the morphology and examine the interface of the particle and surrounding matrix.

We have recently developed a high quality amorphous-Al_2O_3 (a-Al_2O_3) matrix by pulsed laser ablation which is a truly amorphous structures with high band gaps that provides transparency from ~200 nm to ~5 μm. In this work, we report the embedding of metallic particles such as gold (Au) in the a-Al_2O_3 matrix by pulsed laser deposition (PLD). Multiple layer sequences of nanoparticles can be placed within the matrix at specific locations. Changing the matrix can change the dielectric composition surrounding the particles, and high resolution transmission electron microscopy is used to analyze particle morphology and orientation.

EXPERIMENTAL DETAILS

We have used a pulsed laser deposition (PLD) system which comprises a high vacuum system, pulsed excimer laser (λ=248 nm, pulse width=25 ns, and repetition rate=10 Hz) substrate heater and a rotating target holder assembly capable of mounting four targets. The laser energy densities in the range 3.0-4.0 J/cm^2 were used to ablate alumina and gold targets, alternately. Substrates (Si and c-plane sapphire) were heated to $500°C$ during deposition. This was the optimized temperature for a-Al_2O_3 film synthesis. Five alternate sequences of a-Al_2O_3 and Au were deposited in this work. The a-Al_2O_3 film thickness was kept constant while that of Au varied in four such samples. Sample# 1, 2, 3 and 4 had Au layers deposition times as 10 s, 20 s, 40 s and 60 s, respectively. The size of the Au particles was controlled by varying the number of Au monolayers deposited on a-Al_2O_3. The formation of Au crystallites is a self-assembly process governed by Ostwald ripening mechanism. The characterization of these composite films was performed by high resolution transmission electron microscopy (HRTEM) and optical transmission measurements at room temperature.

RESULTS AND DISCUSSION

We studied alumina film without any Au crystallites in it first by high resolution TEM to characterize its microstructure. This was found to be completely amorphous structure without any grain boundaries. The film was found to be optically transparent in 200nm-3.3 μm using a Cary 5E UV-Vis-NIR spectrometer indicating that the band gap was greater than 6.3 eV.

Alumina films with embedded Au Nanoparticles were then characterized. Figure 1 shows a bright field (BF) image of a composite film containing Au crystallites (sample#4). The deposition time of Au-layers in this sample for each sequence was 1 minute. The individual Au grains embedded in alumina matrix are clearly seen. The sizes of these grains were in the range 500-550 Å in this sample. These sizes were in the z-direction (perpendicular to the surface). The electron diffraction (ED) pattern of this sample is seen as inset in Figure 1. The simulated diffraction pattern is also seen in the inset (right corner) which is in agreement with the measured diffraction pattern. The broad halos (figure 1) are from a-Al_2O_3 matrix and the sharp rings are from Au crystallites which is consistent with the cubic-close-packed structure of Au.

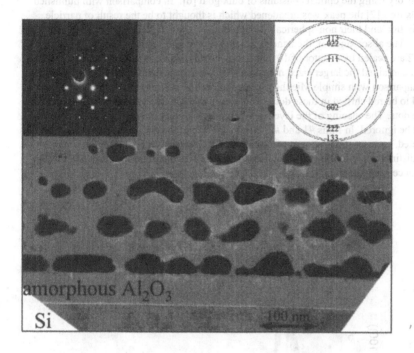

Figure 1: *TEM bright field micrograph and electron diffraction pattern (upper left inset) from the sample#4 in cross-section mode. The upper right inset shows the simulated diffraction pattern of gold.*

The TEM studies done on sample #1 depicted the size of the Au crystallites in the range 300-340 Å. These were the upper and lower bounds and the grain sizes of other two samples were in-between these. The shapes of the Au crystallites was a bit irregular but close to elliptical. We anticipate that changing the deposition paramters to allow more time for the Au atoms to self-assemble will lead to more regular shapes. These were single crystals of Au with no dislocations observed. Our studies on other metals such as silver and nickel have shown that those crystallites tend to be more perfect in shape.

Optical transmission measurements were performed on all 4 samples. The use of sapphire as substrate and amorphous alumina as the matrix permitted measurements out to 200 nm (6.3 eV). In the literature many measurements are limited by the absorption of the substrate and matrix material and data are typically restricted to photon energies less than 4 eV. The use of the sapphire substrate was advantageous since in addition to the plasmon resonance normally seen in gold nanoparticles another broader resonance at higher energies was apparent. The origin of the peak was found to be consistent with Mie theory using the optical constants of bulk gold [6]. In comparison with published calculations [7] the peak was broadened which is thought to be the result of particle distribution and their non-spherical morphology.

The absorbance of the nanoparticles is shown in Figure 2. The data of films #1 and #2 are scaled by a factor of 10 and 4 respectively for readability. Films #4 and #3, which contained the larger sized nanodots, displayed large distinct plasmon resonances. In comparison with simple Mie theory calculations the position of the resonance was found to be red shifted in these dots. This was also attributed to the non-spherical nature of the dots. The spacing of the layers was large enough that interactions between layers could be ignored. In films #1 and #2 the smaller size of the dots damped the resonance as expected, by restricting the mean free path of the electrons. In Film #1 the resonance was very distinct while in Film #2 an increased particle size distribution broadened the resonance substantially.

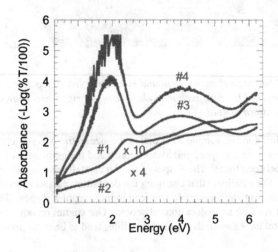

Figure 2: Optical *absorption spectra of all Gold-Alumina samples studied in this work.*

CONCLUSIONS

In conclusion we have found pulsed laser deposition as an effective way to produce metallic nanoparticles in a variety of matrices. Control of deposition parameters, particle size and the dielectric matrix permits control of the plasmon resonance. Further optical studies in conjunction with high-resolution transmission microscopy to characterize particle morphology of several free-electron metals in a variety of dielectric matrices are ongoing.

ACKNOWLEDGMENTS

This work was supported by the NSF Center for advanced materials and smart structures at NC State University.

REFERENCES

1. J.C. Valmalette, L. Lemaire, G.L. Hornyak, J. Dutta, and H. Hofmann, Anal. Mag. **24**, M23 (1996).
2. G. Mie, Ann. Phys. **25**, 377 (1908).
3. C.F. Bohran, and D.R. Huffman, *Absorption and Scattering of Light by Small Particles,* Wiley and Sons, New York, 1983.
4. J.A.A.J. Perenboom, P. Wyder and F. Meier, Phys. Reports, **78**, 173 (1981).
5. B. Palpant, B. Prevel, J. Lemre, E. Cottancin, M. Pellarin, M. Treilleux, A. Perez, J.L. Vialle and M. Broyer, Phys. Rev. **B57**, 1963 (1998).
6. Edward D. Palik, editor, *Handbook of Optical Constants of Solids*, Academic Press, San Diego, 1998.
7. J.A. Creighton, and G.E. Desmond, J. Chem. Soc. Faradey Trans. **87**, 3881 (1991).

Poster Session:
Laser-Solid Interactions
for Materials Processing

Mat. Res. Soc. Symp. Vol. 617 © 2000 Materials Research Society

PLASMA AND DLC FILM CHARACTERISTICS FROM PULSED LASER ABLATION OF SINGLE CRYSTAL GRAPHITE AND AMORPHOUS CARBON: A COMPARATIVE STUDY EMPLOYING ELECTROSTATIC PROBE MEASUREMENTS

R. M. Mayo, J. W. Newman
Department of Nuclear Engineering, North Carolina State University,
Raleigh, NC 27695-7909

A. Sharma, Y. Yamagata,* J. Narayan
Department of Materials Science and Engineering, North Carolina State University,
Raleigh, NC 27695-7907
*Department of Electrical and Computer Engineering, Kumamoto University, Kumamoto 860-8555, Japan

ABSTRACT

In an ongoing effort to investigate plasma plume features yielding high quality DLC films, we have applied plasma plume diagnosis and film characterization to examine plume character distinction from KrF laser ablation of both amorphous carbon (a-C) and single crystal graphite (SCG) targets. The advancing plasma plume produced by these structurally different targets are observed to possess quantitatively similar total heavy particle inventory, ionized fraction, and electron thermal content, yet quite different ion kinetic energy, plume profile, C_2 formation mechanism, and concentration of complex molecules. All data support the conclusion that the SCG target plasma plume is populated with heavier, more complex molecules than those in a-C which have been shown to be predominantly comprised of C and C^+ under vacuum conditions with the addition of C_2 at high fill pressure. Significantly smaller plume profile peaking factor, less energetic and slightly lower temperature plume conditions, laser energy (E_l) dependent plume peaking, harder films produced at lower E_l, strongly heterogeneous films, and lesser plume energy attenuation in high pressure background fill in SCG target plumes all support the conclusion of more massive plume species in SCG target plumes. Energy balance estimates indicate that ion kinetic energy dominates and that SCG target ablation liberates about twice the number of ^{12}C atoms per unit E_l.

INTRODUCTION

Development of the Pulsed Laser Deposition (PLD) technique has enabled the growth of a wide variety of novel thin film structures[1-5] at low processing temperatures with high crystalline quality and low contamination levels while preserving stoichiometry. As the average energy of pulse laser ablated species is much greater than kT, it is possible to form metastable phases. Among the more important films developed in this way include multilayer and multicomponent stoichiometric composites, epitaxial superconducting thin film heterostructures, and diamond-like carbon (DLC)[2,5,6]. The latter constitutes an important area of recent and vigorous research activity[5-7]. The value of DLC films emanates from their electrical, optical, and mechanical properties approaching those of diamond[5] such as density and hardness, high electrical resistivity, low friction coefficient, optical transparency, and chemical inertness. The complex and interdependent nature of target evaporation, plasma plume development, and film deposition depend strongly on a number of process variables and mechanisms. As the plasma state is the intermediary in the PLD process, it plays a crucial role in determining eventual film properties. Thereby, we apply plasma diagnosis to elucidate plasma and flow dynamics, and plume atomic and molecular processes in these energetic plasma plumes.

Recent work indicates strong influence of a small number of plasma parameters; ion kinetic energy, ion fraction, and plasma plume molecular concentration, as contributing to DLC film quality. Ion kinetic energy (KE_i) has been demonstrated to positively influence DLC film character[6,8-10], ultimately through direct correlation with sp^3 bonding fraction where some data suggest an optimal KE_i near ~ 90 eV for maximizing sp^3 ratio. Our recently established experimental effort[6,7] supports the importance of ion directed energy through correlation of film bonding characteristics as measured by Raman spectroscopy, with electrostatic ion probe measurement of plume advancement. We have also carefully described ion inventory in the propagating ablated plasma plume and have shown an ion fraction of ~ 10 % [8–10]. Our film analysis further shows this ion inventory in agreement with the deposited fraction[6]. Combined with the demonstrated importance of KE_i, this result suggests that it is the flux of high energy ions that is principally responsible for DLC formation. In addition, we have shown[6,7] that the conditions concomitant with high molecular content (esp. C_2) necessarily degrade DLC film character.

We present a systematic, comparative study of plasma plume character generated by KrF pulse laser ablation of amorphous carbon (a-C) and single crystal graphite (SCG) targets by employing electrostatic ion probe, optical spectroscopy, Raman spectroscopy, and nano-hardness measurements to assess plume and film properties. Plasma and flow parameters are presented and compared for the two targets showing notably slower propagation for the SCG plume. From the electrostatic probe data, plume morphology is inferred showing a much more forward directed flow for the a-C target plume. It is further inferred from all the acquired data, that plasma plumes from SCG targets consist of heavier particles (more complex molecules) on average. Since SCG possesses strong in-plane bonds and weaker bonds between planes, it is suggested that complex species with greater mass[11–14] may be formed to minimize energy. Pulsed laser ablation of SCG may have the potential to produce carbon composites enriched in such complex species.

EXPERIMENT

The PLE experiment facility at the North Carolina State University has been described previously[2,6]. The experiment consists of a stainless steel vacuum chamber supporting a target and substrate, excimer laser, and plume diagnostics. The chamber is evacuated to $\lesssim 10^{-7}$ Torr by a turbo-molecular pump. In the experiments described here, the target consists of either 99.995% polycrystalline graphite (a-C) in the form of a 25 mm diameter disk with a mass of approximately 3.6 g, or single crystal graphite (SCG) in a 10 mm square with a mass of approximately 0.36 g. The target is rotated at ~ 6 rev/min to prevent pitting. During film deposition the room temperature Si (100) substrate is mounted parallel to the target at 3.5–6.0 cm separation. Target ablation is provided by a Lambda Physik KrF (248 nm) excimer laser. The beam path is incident on the target at 45° and delivers 2.0–4.5 J/cm^2. The laser energy to fluence calibration is $\mathcal{E}(\text{J/cm}^2) \simeq 11.64\,\text{cm}^{-2}\,E_l(\text{J}) + 0.427\,\text{J/cm}^2$ for laser energy in the range $0.220 \leq E_l \leq 0.345$ J. With a ~ 45 ns [25 ns full width at half maximum (FWHM)] pulse width, this gives an average power density of 80–200 MW/cm^2 on target. Under these near-threshold conditions a luminous, high temperature (0.9–3.0 eV) plasma plume is ejected in an elongated shape perpendicular to the target at high speed. The plasma particle density in the isothermal layer immediately adjacent to the target is model estimated[4] to be on the order of 10^{19}–10^{21} cm^{-3}. The elongated plume shape suggests highly forward directed flow, consistent with film depth profiles[15] indicating a $\cos^m \theta$ distribution, where $8 \leq m \leq 12$ and θ is measured from the normal to the a-C target at the laser interaction point. Plasma plume profiles in the SCG target case are marked by smaller peaking parameter, m, indicating a less forward directed propagation. High kinetic energy of the ion ejecta, $KE_i \gg kT$, supports the *space charge acceleration model*[16].

An Optical Multichannel Analyzer (OMA)[7] and a triple Langmuir probe[17] constitute the principal plume plasma diagnostic tools. Visible and near UV spectra are recorded with the OMA to observe relative trends in C_2, C, and C$^+$ emission as conditions are systematically varied. Electrostatic measurements of electron density and temperature are performed with a cylindrical (5 mm long, 10 mil dia.), tungsten wire, triple probe. The probe axis is positioned normal to the target, but slightly (~ 5 mm) off axis from the target center-line at the laser interaction point. A sliding seal allows probe positioning without vacuum break. Time resolved (2–5×10^6 sample/s) probe data are recorded at 1, 3, and 5 cm from the target. The bias potential between the negative polarity (with respect to plasma potential) probe tip and positive polarity probe tip is typically -7 to -30 V to ensure ion saturation.

Sample Langmuir probe data are shown in Fig. 1 for 270 mJ (~ 3.6 J/cm^2) KrF laser pulses in vacuum with laser firing time at $t \sim 1.4$ μs. The center-line electron density displayed in Fig. 1 is computed assuming C^+ ions dominate the ionic species concentration

$$n_e = \frac{i_{sat}}{0.61Ae\sqrt{kT_e/M} + A_\perp eu} \tag{1}$$

where i_{sat} is the ion saturation current to the negatively biased probe tip, M is the C^+ ion mass, $A = 2\pi a l \simeq 4 \times 10^{-6}$ m^2 is the probe cylindrical surface area, $A_\perp$ is the effective probe tip area perpendicular to the bulk ion flow direction, and u is the ion flow speed in the plume. The first term in the denominator of Eq. (1) represents the thermal ion flux (per unit density) to the probe satisfying the Bohm criterion[17]. The second term provides a correction due to supersonic ion flow[18] intersecting the ion saturation probe tip and off axis positioning of the probe. Since the ion drift speed $u \gg C_s$, where $C_s = \sqrt{kT_e/M}$ is the ion acoustic speed, the correction can be significant even as $A_\perp < A$. The ion drift speed u and $KE_i = \frac{1}{2}Mu^2$ are estimated from data like that in Fig. 1 by considering the leading edge time difference of the i_{sat} signal between two consecutive probe positions. The drift time between probe positions displaced by 2 cm in the a-C plume at 220 mJ (Fig. 1) is estimated at ~ 0.4 μs, giving a steady drift speed of $\sim 5.0 \times 10^4$ m/s and $KE_i \sim 156$ eV. Greater separation times between subsequent probe signals are observed in SCG plume density. At the same laser energy, ion kinetic energy estimates in SCG are $KE_i^{SCG} \sim (\frac{1}{4} - \frac{1}{2}) KE_i^{a-C}$.

RESULTS

Under high vacuum conditions ($\sim 10^{-7}$ Torr), electron temperatures (T_e) for the SCG case are consistently 10–30% higher than that for the a-C target at the same location, indicating a greater freeze out temperature in the SCG case. It is expected that radiation is responsible for the decrease in T_e with flow propagation away from the target in both cases. Normalized plume center-line density data are summarized in Fig. 2 for both a-C and SCG targets by referencing each individual data point to the maximum density for either target at each axial position. The thin solid lines separate the a-C from the SCG data. Here it can be seen that a-C plume densities are consistently larger that those in SCG plumes. Closest to the target, at the 1 cm position, the difference is 10–12%, consistent with target mass loss measurements of total heavy particle inventory[6] indicating a loss of 0.36 μg/pulse or $\sim 1.8 \times 10^{16}$ carbon atoms/pulse in a-C and 0.316 μg/pulse or $\sim 1.6 \times 10^{16}$ carbon atoms/pulse in SCG at 240 mJ. The SCG data show much greater density spread than for the case of a-C. This spread is also systematic in E_l, so that at each axial location the SCG target plume density increases with E_l.

The significantly lower center-line density (n_{e_o}) for downstream positions (> 1 cm) and speed of propagation for the SCG plume over that for a-C, suggest substantial differences in plume profile. From our heavy particle density model described earlier[6], we have

$$n(r, \theta) = \frac{m+1}{4\pi\Delta} N \left(\frac{\cos^m \theta}{r^2} \right) \tag{2}$$

where N is the heavy particle inventory, m is the profile peaking parameter, and Δ is the azimuthally symmetric plume thickness, typically ~ 5 cm for a-C and ~ 3 cm for SCG. Preserving the non-interacting plume r^{-2} profile dependence results in estimates of $m \sim 9$ for a-C and $m \sim 3$ for SCG. This result is consistent with independent theoretical models for plume non-uniformities[19] that demonstrate stronger forward peaking as mass dependent and should result from lower mass plumes propagating at higher flow velocity and at lower temperature, suggesting greater average mass particles in the plasma plume generated from ablation of SCG targets. The systematic E_l dependence in the normalized density for the SCG target in Fig. 2 shows monotonic increase for peaking factor in the range $2 \lesssim m \lesssim 4$ as E_l increases from 220 mJ to 345 mJ. Ion fraction estimates based on our heavy particle density model (Eq.(2)) for the a-C target case have been presented earlier[6] and are found to consistently reside in the range 10–15% in good agreement with the findings of others[9]. Proceeding likewise for the SCG target case with $m = 3$ we find a similar range of ion fraction estimates.

Film nano-hardness data indicate harder films from the SCG target at lower E_l, suggesting that increased E_l tends to break-up or otherwise prevent the formation of more complex carbon molecules.

These new structures may be responsible for both harder films and heavier, slower, less forward peaked plumes at lower E_l in SCG. The micro-Raman data of Fig. 3 support this assertion. Here we show Raman spectra (1150–1700 cm^{-1}) for three sections in close proximity of a film formed by deposition of an ablated SCG target in vacuum at 250 mJ. Clearly evident here are adjacent regions of dramatically different film character indicating strongly heterogeneous films are formed during SCG ablation in vacuum. By contrast, a-C target ablation produces films with monotonically increasing hardness as E_l is increased further supporting the importance of energetic atomic carbon and/or ions in promoting sp^3 bond formation.

The principal differences between plasma plumes produced by ablation of the two structurally different targets are plume profile peaking parameter, propagation speed, and complex molecule concentration. Since ablated particle inventory, temperature, and ion fraction are nearly identical, we expect thermal energy content, radiated energy, and ionization losses to be similar for the two targets. A simple energy balance argument reveals that reflection may account for the lesser absorbed laser energy for the SCG target. With thermal and ion kinetic content being the predominant contributions to energy balance, they must account for the bulk of the absorbed laser energy $(1 - R)E_l$ such that

$$(1 - R)E_l = (1 + 2f_i)NkT_e + NKE_i \qquad (3)$$

where R is the laser light reflection coefficient, N is the total heavy particle inventory, and it is assumed that all species (ions, electrons, and neutral atoms) are in thermal equilibrium at T_e. Applying expression (3) in ratio for the two target cases proves much less sensitive to the absolute energy balance components. With $f_i \sim 0.1$, $T_e^{SCG} \sim 1.15T_e^{a-C}$, $N_{SCG} \sim 0.9N_{a-C}$, and $KE_i^{SCG} \sim \frac{1}{2}KE_i^{a-C}$, we find $(1 - R_{SCG})/(1 - R_{a-C}) \sim 0.46$. The black a-C target has $R_{a-C} \sim 0$, so that we estimate $R_{SCG} \sim 0.54$. Then about half the incident laser energy is then reflected in the SCG case. Per unit absorbed laser energy, the SCG target liberates approximately twice as many ^{12}C atoms, as expected by the lesser bonding energy and the potential to be released as more complex molecules.

Ablation in high pressure N$_2$ fill has been studied for its influence on DLC film properties[6,7], plasma plume dynamics[6,9], and fullerene and nanotube formation[14]. Figure 4 shows the raw i_{sat} signals for the a-C at 220 mJ and 200 mTorr N$_2$ background. As in high vacuum conditions, the SCG plumes are consistently slower at low pressure (0–300 mTorr), reflecting the less energetic nature and the presence of heavier particles. Consistent with earlier results[6], a second density peak (plume splitting) is clearly visible in these data for the 1 cm signal. Second peaks are present at 3 and 5 cm, but are convolved with the broadened primary peak (cf: Fig. 1). In the SCG case, the second peak is more clearly separated from the primary density spike especially at the 1 cm position, another indication of a more lethargic SCG plume. Mass diffusivity estimates[6] suggested a dominating contribution from C$_2^+$ in the a-C plume secondary peak. Slower features in the SCG data suggest the presence of heavier, more complex molecules than C$_2$. The effect of high pressure N$_2$ fill is also to reduce and broaden the i_{sat} signals.

Figure 5 shows the second peak magnitude at the 1 cm position normalized to the primary i_{sat} signal as a function of N$_2$ fill pressure. In both the a-C and SCG cases the ratio monotonically increases with pressure in agreement with optical C$_2$ emission. This ratio, however, increases much more rapidly for the a-C case, eventually exceeding unity beyond $\sim$ 300 mTorr. This suggests very different molecular formation mechanism(s). Since no C$_2$ emission is seen at high vacuum conditions in the a-C plume[7], we infer that C$_2$ is formed within the a-C plume at high pressure. It was shown earlier[6] that these conditions are concomitant with degraded DLC film quality indicating C$_2$ is unimportant in DLC formation. In the SCG plume, C$_2$ emission is observed at all pressure conditions, indicating the presence of molecular carbon generated at the target, where fragmentation may also play a role in generating C$_2$ from more complex molecules. The presence of an energetic primary SCG peak indicates that there exists substantial concentration of C and C$^+$ mixed with heavier molecules in the SCG plume as is expected from the heterogeneous film structure discussed earlier (Fig. 3).

The effect of pressure on plume propagation is summarized in Fig. 6 where leading edge KE$_i$, determined from time-of-flight between the 1 and 3 cm positions, is shown. The retarding influence of the background fill is evident in both a-C and SCG cases as KE$_i$ is substantially reduced. Since KE$_i^{a-C}$ is always greater than KE$_i^{SCG}$ in vacuum, there is a cross-over region indicated by the shaded area in Fig. 6, wherein the initially more energetic a-C plume becomes more strongly attenuated. Cross-over implies the SCG plume consists of more massive particles.

CONCLUSIONS

We have applied plasma plume diagnosis and film characterization in an effort to investigate plasma plume features yielding high quality DLC films. We have further examined plume character distinction upon ablation of both a-C and SCG targets. The advancing plasma plume produced by these structurally different targets are observed to possess quantitatively similar particle inventory, ion fraction, and electron thermal content, yet quite different ion kinetic energy, plume profile, C_2 formation mechanism, and concentration of complex molecules. Plume electron temperatures are found to reside in the range 1–3 eV with those in SCG plumes $\sim$ 10–30 % greater than a-C at all positions. For both cases, we find T_e drop off with position away from the target. By contrast, electron density is found to be $\sim$ 10–12 % lower near the target in SCG than a-C plumes consistent with mass loss inventory measurements. Applying appropriate profile peaking factors consistent with r^{-2} dependence and theoretical predictions, we infer ion fractions in the range $\sim$ 10–15 % for both target cases.

All data support the conclusion that the SCG target plasma plume is populated with heavier, more complex molecules than those in a-C which have been shown to be predominantly comprised of C and C^+ under vacuum conditions with the addition of C_2 at high fill pressure. Significantly smaller plume profile peaking factor, less energetic and slightly lower temperature plume conditions, laser energy (E_l) dependent plume peaking, harder films produced at lower E_l, strongly heterogeneous films, and lesser plume energy attenuation in high pressure background fill in SCG target plumes all support the conclusion of more massive plume species in SCG target plumes. Energy balance estimates indicate that ion kinetic energy dominates and that SCG target ablation liberates about twice the number of ^{12}C atoms per E_l.

ACKNOWLEDGEMENTS

This work was performed under the auspices of the National Science Foundation, Center for Advanced Materials and Smart Structures.

REFERENCES

[1] J. Narayan et al., Appl. Phys. Lett. **51**, 1845 (1987).
[2] J. Krishnaswamy et al., Appl. Phys. Lett. **54**, 245 (1989).
[3] J. P. Zheng et al., Appl. Phys. Lett. **54**, 280 (1989).
[4] R. K. Singh and J. Narayan, Phys. Rev. B **41**, 8843 (1990).
[5] D. H. Lowndes et al., Science **273**, 898 (1996).
[6] R. M. Mayo et al., J. Appl. Phys. **86**, 2865 (1999).
[7] Y. Yamagata et al., J. Appl. Phys. **86**, 4154 (1999).
[8] P. T. Murray and D. T. Peeler, J. Elec. Mater. **23**, 855 (1994).
[9] D. B. Geohegan, in *Pulsed Laser Deposition of Thin Films*, edited by D. B. Chrisey and G. K. Hubler (John Wiley and Sons, Inc., New York, 1994), chapter 5, p. 115.
[10] V. I. Merkulov et al., Appl. Phys. Lett. **73**, 2591 (1998).
[11] P. S. R. Prasad et al., Phys. Stat. Sol. A **139**, K1 (1993).
[12] H. Koinuma et al., Fullerene Sci. Technol. **4**, 599 (1996).
[13] L. Laska et al., Carbon **34**, 363 (1996).
[14] Y. Zhang et al., Appl. Phys. Lett. **73**, 3827 (1998).
[15] R. K. Singh et al., J. Appl. Phys. **68**, 233 (1990).
[16] H. Opower and W. Press, Z. Nat. Forsch. **21a**, 344 (1966).
[17] S. L. Chen and T. Sekiguchi, J. Appl. Phys. **36**, 2363 (1965).
[18] S. B. Segall and D. W. Koopman, Phys. Plasmas **16**, 1149 (1973).
[19] K. L. Saenger, J. Appl. Phys. **70**, 5629 (1991).

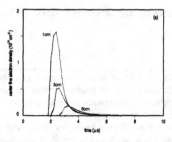

Figure 1: Electron density measured by the triple Langmuir probe for KrF laser pulses in vacuum at 1, 3, and 5 cm from the a-C target at 270 mJ (~ 3.6 J/cm^2). The time base is referenced to the laser trigger pulse time at $t = 0$. The laser pulse arrives at ~ 1.4 μs in all figures shown.

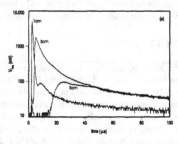

Figure 4: Electrostatic probe ion saturation signals as raw digitizer traces (mV) for KrF laser pulses at 200 mJ and 200 mTorr N$_2$ background at 1, 3, and 5 cm from the a-C target.

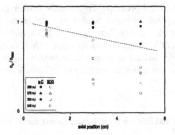

Figure 2: Electron density as a function of axial position for both a-C (filled) and SCG (open) target plumes in vacuum normalized to the peak density at each position.

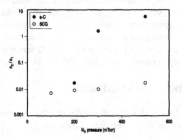

Figure 5: Second i_{sat} peak magnitude relative to that of the primary C$^+$ peak.

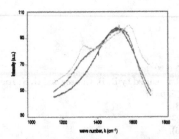

Figure 3: Micro-Raman data from deposited film formed by ablation of an SCG target at 250 mJ in vacuum. Film character is strongly heterogeneous showing high quality DLC character (black curve) in close proximity to a softer film (light grey curve).

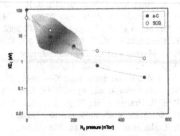

Figure 6: Ion kinetic energy as a function of N$_2$ pressure for both a-C and SCG generated plasma plumes. The shaded area indicates the cross-over region.

Mat. Res. Soc. Symp. Vol. 617 © 2000 Materials Research Society

Improvement of Cavitation Erosion Resistance and Corrosion Resistance of Brass by Laser Surface Modification

K. F. Tam[1], F. T. Cheng[1], H. C. Man[2]
[1]Department of Applied Physics,
[2]Department of Manufacturing Engineering,
The Hong Kong Polytechnic University, Hong Kong, People's Republic of China

ABSTRACT

Laser surface modification of brass (Cu-38Zn-1.5Pb) using AlSiFe and NiCrSiB alloy was achieved by using a 2kW continuous wave Nd-YAG laser with the aim of improving the cavitation erosion resistance and corrosion resistance. The alloying powder was preplaced on the brass substrate by thermal spraying to a thickness of 350μm, followed by laser beam scanning to effect melting, mixing and alloying. A modified surface was achieved by overlapping of adjacent tracks. The cavitation erosion resistance and the anodic polarization characteristics of the laser surface modified specimens in 3.5% NaCl solution at 23°C were studied by means of a 20kHz ultrasonic vibrator at a peak to peak amplitude of 60μm and a potentiostat respectively. The cavitation erosion resistance of the specimens modified with AlSiFe and NiCrSiB was improved by a factor of 3 and 7 respectively, compared with that of the brass substrate. Potentiodynamic test, however, indicated that the corrosion resistance of specimens modified with AlSiFe deteriorated, as reflected by a shift of the polarization curve towards higher current densities. On the other hand, the corrosion resistance of specimens modified with NiCrSiB was significantly improved, as evidenced by the presence of a passive region (from −175 mV to −112 mV) and a reduction in the anodic current density by at least an order of magnitude compared with the substrate at the same anodic potential. The hardness profile and the compositional profile were measured using a Vickers hardness tester and EDX respectively. The microstructure and the surface morphology of the specimens were investigated with the aid of SEM and optical microscopy.

INTRODUCTION

Damage due to cavitation erosion is a common problem in engineering parts exposed to a fast flowing or vibrating liquid, where cavities or bubbles are generated and then collapse due to pressure fluctuations. The shock wave or micro-jet generated during bubble collapse repeatedly exert stress pulses on the parts near by, causing fatigue failure and erosion[1].

Brass, a copper-zinc alloy with a zinc content varying from 10% to 40%, is a major copper alloy in engineering applications. The addition of zinc improves the mechanical strength and reduces the cost of the alloy, but at the same time lowers its corrosion resistance[2]. In fact, owing to the large difference in electrochemical activity of zinc and copper, brass with zinc content >15% is susceptible to dezincification. Brass is also less resistant to cavitation erosion than the more expensive copper alloys such as cupro-nickel. The cavitation erosion resistance and corrosion resistance of brass may be improved by laser surface modification, a process which retains the bulk properties of brass, uses only a small amount of additional material for treatment, and produces surface layers of unique properties[3,4]. However, owing to the high reflectivity and high thermal conductivity of copper and its alloys, laser energy absorption and hence laser treatment of copper alloys impose a certain degree of difficulties in the process[5]. The present study aims at surface modifying a high-zinc brass (Cu-38Zn-1.5Pb) with NiCrSiB and AlSiFe using a 2kW CW Nd-YAG laser. The cavitation erosion and corrosion resistance of the modified layer in 3.5% NaCl solution at 23°C will also be investigated.

EXPERIMENTAL DETAILS

The nominal chemical composition of the as-received brass specimen, in the form of rectangular bars, is shown in Table I. Specimens of $30 \times 20 \times 8$mm were cut from the rectangular bar. Before laser surface modification, NiCrSiB or AlSiFe powder of nominal chemical composition shown in Table I was pre-placed on the specimens by flame spraying, which was achieved by using an acetylene thermal spraying gun. A digital micrometer was used to measure the average thickness of the sprayed layer.

A 2kW CW Nd-YAG laser (LUMONICS, JK Multiwave, Model MW2000) was employed for laser surface modification. The diameter of the Nd-YAG laser spot on the substrate was set at 3mm approximately. The laser power used was 1.0 kW and 1.2kW for NiCrSiB and AlSiFe respectively, and the corresponding scanning speed was 5mm/s and 20mm/s. The coverage of the whole surface of specimens was achieved by overlapping 40% of single melt tracks. Argon was used as shielding gas with a flowrate of 10 l/min.

Following laser treatment, the specimens were sectioned perpendicular to the melt track, polished with 1/4μm diamond paste and etched with acidic ferric chloride solution (5cc FeCl3, 10cc HCl, 100cc H2O). The microstructure and chemical composition in the laser treated specimens were analysed by optical microscopy (OM), scanning electron microscopy (SEM) and energy dispersive spectroscopy (EDX). Microhardness was obtained using a Vickers hardness tester, the load applied being 200 g and the loading time being 15 seconds. The microhardness at different points on the cross section was measured to obtain a hardness profile.

An ultrasonic induced cavitation facility was employed to perform the cavitation erosion test conforming to ASTM Standard G32-92[6], with minor modification in the method of mounting the specimen. The treated specimen was held stationary at a distance 1 mm below the horn. Such a method has also been employed by some other researchers[7]. The vibratory frequency used was 20 kHz and the peak to peak amplitude was set at 60 μm by a dial indicator. The solution was contained in a 1000 ml open container and maintained at constant temperature by a circulation system with controllable temperature water bath. The laser treated specimens were subjected to a series of cavitation erosion tests in 3.5% NaCl solution at 23°C. Each cavitation erosion test was completed after 4 hours, which included eight intermittent periods each of half an hour, and the specimen was weighed to obtain the mass loss every half an hour. The erosion loss of materials was expressed in terms of the mean depth of penetration (MDP) and, the mean depth of penetration rate (MDPR) as calculated by the following equations:

Table I Nominal chemical compositions of brass substrate, NiCrSiB and AlSiFe

	Cu	Zn	Pb	Ni	Al	Cr	Si	B	Fe	C
Brass (Cu-Zn)	Bal.	38	1.5	-	-	-	-	-	-	-
Ni-Cr-Si-B powder (~100μm)	-	-	-	Bal.	-	16.5	3.5	3.8	15.5	<1
Al-Si-Fe powder (~100μm)	-	-	-	-	Bal.	-	19.5	-	7.5	-

$$MDP(\mu m) = \frac{\Delta W}{10\rho A} \qquad\qquad [1]$$

$$MDPR(\mu m\,h^{-1}) = \frac{\Delta W}{10\rho A \Delta t} \qquad\qquad [2]$$

where ΔW the weight loss in each time interval in mg
Δt the time interval in h
A the surface area of the specimen in cm^2
ρ the estimated density of the modified layer in g cm^{-3}

The cavitation erosion resistance (R_e) is defined as the reciprocal of MDPR

$$R_e(h\,\mu m^{-1}) = (MDPR)^{-1} \qquad\qquad [3]$$

The laser treated specimens were mounted by hot curing powder for corrosion testing, leaving an exposed area of about 10×10 mm^2 for electrochemical investigations. Adhesive epoxy was applied to cover the specimen-plastic interface to prevent any leakage and any adverse crevice effect. A potentiostat (EG&G PARC 273 corrosion system) was used to study the corrosion characteristics of the untreated and treated specimens by potentiodynamic polarization technique according to ASTM Standard G5-92[8]. The test solution was 3.5% NaCl solution at temperature 23°C, and open to air. A delay time of 10 minutes and a potential sweep rate of 1mV/s were used, starting from –100mV below the corrosion potential. All potentials were measured with respect to the Saturated Calomel Electrode (SCE, 0.244V versus Standard Hydrogen Electrode (SHE) at 25°C) as the reference electrode. Two parallel graphite rods served as the counter electrode for current measurement. Corrosion parameters including the corrosion potential *Ecorr*, pitting potential *Epit* and corrosion current density *Icorr* were extracted from the polarisation curves.

RESULTS AND DISCUSSIONS

Microstructural and metallographic analysis

The flame-sprayed coating was about 350 μm thick for both types of powder. The coating was uneven and porous. Under laser irradiation, the alloy powder and the underneath brass substrate were melted and intermixed, and rapidly solidified to form a modified layer. The alloyed layer was free from pore and cracks and bonded to the brass substrate by metallugurical fusion bonding.

Fig.1 shows the hardness profiles of the modified surfaces. It is obvious from the profile that the hardness was very significantly increased in the case of NiCrSiB (with an average value of 378Hv in the modified layer), while for AlSiFe the increase was only minimal (with an average value of 148Hv). The results of hardness measurement were consistent with the compositional profiles of the modified layers shown in Fig.2. In the layer modified with NiCrSiB, the alloying elements were homogeneously distributed down to a thickness 350μm. However, only small amounts of the alloying elements were present in a surface layer of about 20μm thickness in the case of AlSiFe. The small increase in hardness in the latter case might result from the presence of Al and also from grain refinement of the substrate in laser treatment.

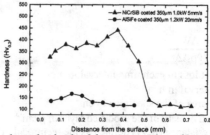

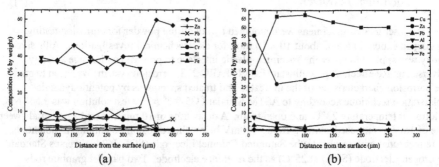

Fig. 1 Hardness profiles along the depth of the cross section of laser surface modified specimens

Fig. 2 Chemical compositional profiles of laser modified specimens (a) NiCrSiB/brass (b) AlSiFe/brass

Cavitation erosion of laser surface modified brass

Fig.3 shows the cumulative MDP of the as-received brass specimen and the laser modified specimens cavitated in 3.5% NaCl solution at 23°C as a function of time. The normalised R_e values with respect to that of the as-received brass specimen are shown in Table II. The ranking of the specimen in cavitation erosion resistance is NiCrSiB/brass > AlSiFe/brass > As-received brass. The R_e of NiCrSiB/brass specimen was improved by about 7 times, while that for the AlSiFe/brass specimen was improved by 3 times. The improvement in cavitation resistance was consistent with the increase in hardness.

Electrochemical corrosion of laser surface modified brass

The potentiodynamic polarization curves of the as-received brass and the laser modified specimens in 3.5% NaCl solution at 23°C were shown in Fig.4. The corrosion potentials of the various specimens were very close to the value of the as-received specimen. For the AlSiFe/brass specimen, the corrosion resistance deteriorated as indicated by a shift of the polarization curve towards higher current densities. On the other hand, the resistance to pitting corrosion for the NiCrSiB/brass specimen was significantly improved. The pitting potential was shifted in the noble direction by an amount of about 63mV (from −175mV to −112mV), while the current density was at least an order lower than that of the as-received specimen at the same anodic potential.

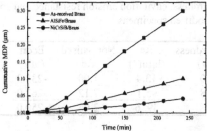

Fig. 3 Cumulative mean depth of penetration (MDP) as a function of time for the as-received and laser surface modified specimens eroded in 3.5% NaCl solution at 23°C

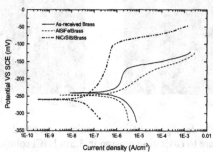

Fig. 4 Potentiodynamic polarization curves of the as-received brass and the laser modified specimens in 3.5% NaCl solution at 23°C

Cavitation damage mechanism

The appearances of the damaged surface of specimen NiCrSiB/brass and AlSiFe/brass after 4 hours of cavitation test were shown in Fig.5. The surface was roughened after the 4-hour test, indicating that material was eroded from the surface by ductile fracture.

CONCLUSIONS

Brass specimens (Cu-38Zn-1.5Pb) were laser surface alloyed with NiCrSiB and AlSiFe powder using a laser power of 1.0 kW and 1.2 kW, and a scanning speed of 5mm/s and 20mm/s respectively (beam size of 3mm diameter). For the NiCrSiB/brass specimen, the microhardness of the modified layer increased (from 110 Hv to 378Hv), while the improvement in the AlSiFe/brass specimen was smaller (from 110Hv to 148Hv). However, the cavitation erosion resistance of the two modified specimens was improved by a factor of 7 times (NiCrSiB) and 3 times (AlSiFe) compared with that of the substrate. Moreover, the corrosion resistance of the NiCrSiB/brass specimen in 3.5% NaCl solution was significantly improved, as evidenced by an increase of 63mV in the pitting potential and a reduction of the corrosion current density by more than an order of magnitude. For the modified layer using AlSiFe, the corrosion resistance deteriorated as indicated by a shift of the polarization curve towards higher current densities.

Table II The average hardness, cavitation erosion resistance and corrosion parameters of as-received and laser surface modified specimens

Specimens	Ave. hardness (Hv)	Re (hμm⁻¹)	Normalised Re	Ecorr (mV)	Icorr (μA/cm²)	Epit (mV)
As-received Brass	110	13.4	1.00	-238.3	1.00	-175
NiCrSiB/brass	378	95.1	7.12	-262.0	0.09	-112
AlSiFe/brass	148	39. 6	2.96	-250.0	1.00	--

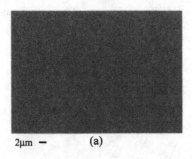 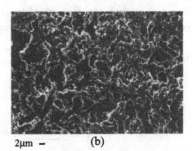

2μm —   (a)   2μm —   (b)

Fig. 5 Appearance of the damaged surface of specimen modified with (a) NiCrSiB and (b) AlSiFe after 4 hours exposure to cavitation erosion in 3.5% NaCl solution

ACKNOWLEDGEMENT

The authors would like to acknowledge the Research Committee of the Hong Kong Polytechnic University for the provision of a research grant (No. G-V710).

REFERENCES

1. A. Karimi, J. L. Martin, *International Metals Reviews*, **31**, 1-26 (1986).
2. N. W. Polan, "Corrosion of Copper and Copper Alloys" in: *ASM Handbook*, **Vol. 13**, Corrosion, ASM International, USA, (1992).
3. W. M. Steen, in: C. W. Draper, P. Mazzoldi (Eds), *Laser Surface Treatment of Metals*, 387-396, Martinus Nijhoff Publishers, Dodrechent, (1996).
4. B. L. Mordike, in: R. W. Cahn, P. Haasen, E. J. Kramer (Eds), *Materials Science and Technology: Processing of Metals and Alloys*, **Vol. 15 VCH**, New York, 111-136 (1991).
5. G.Dehm, et.al., *Wear*, **225-229**, 18-26 (1999).
6. ASTM Standard G32-92, Standard Test Method for Cavitation Erosion Using Vibratory Apparatus, in: *Annual Book of ASTM Standards*, **Vol. 03.02**, ASTM, Philadelphia, (1992).
7. B. Vyas, I. L. H. Hansson, *Corrosion Science*, **30**, 8-9, 261-270 (1990).
8. ASTM Standard G5-94, Standard Reference Test Method for Making Potentiostatic and Potentiodynamic Anodic Polarization Measurements, in: *Annual Book of ASTM Standards*, **Vol. 03.02**, ASTM, Philadelphia, (1994).

Mat. Res. Soc. Symp. Vol. 617 © 2000 Materials Research Society

Ablation-induced stresses in fused silica by 157-nm F$_2$-laser irradiation

Igor A. Konovalov and Peter R. Herman
Department of Electrical and Computer Engineering
University of Toronto
10 King's College Road, Toronto, ON, Canada, M5S 3G4

ABSTRACT

The F$_2$ laser is a promising source for direct etching of microstructures and the precise shaping of optical-grade surfaces on wide bandgap materials such as fused silica. We report here on residual tensile stresses induced in fused silica (Corning 7940, UV grade) by 157-nm laser ablation. Plastic strain of 160-mm thick rectangular strips, monitored with an optical interferometric microscope, revealed the presence of residual tensile stresses in the near-ablated surface. HF chemical thinning of the sample showed the thickness of ablation-affected layer provoking strain was ~275 nm, a value independent of laser fluence (1.9-4.7 J/cm^2) and scanning speed (94 - 220 µm/s). A near-surface mean residual tensile stress of ~80 MPa was inferred from a thin film-substrate approximation.

INTRODUCTION

Thin-film devices and systems consisting of dissimilar materials or layers frequently exhibit substantial residual stresses as noted in numerous studies of thin films [1]. Large residual stresses can also develop during surface treatment involving ion, electron, or laser beams. While such treatments offer improved surface properties, the residual stresses become increasingly important in microsystems where feature sizes are shrinking to the dimensions of the affected surface layer. Residual stresses can hinder the operation of microelectronic, photonic or biological components and can lead to bending or buckling of micro-electromechanical devices (MEMs).

In laser material processing, a field of widespread commercial significance, residual stresses arise in the shaping and machining of materials and in the texturing and modification of surfaces. Laser-induced residual stresses can grossly affect the macroscopic properties of materials, when using ultraviolet lasers such as excimers to define feature sizes on the scale of 100's of nanometers. This paper examines the stresses formed by laser ablation of fused silica, a high-quality material widely used in optical and photonic applications.

Laser damage and laser ablation of fused silica has been extensively studied using conventional sources such as CO$_2$, YAG, visible, and excimer [2-7]. Several kinds of damage related to residual stresses have been noted for fluences below the ablation threshold. Material densification and concomitant refractive index changes lead to residual tensile stresses in irradiated glasses [13]. Another form of damage includes the generation of surface or internal cracks, and the manifestation of residual tensile-stress fields around the crack [14-15]. Microcrack formation is common for fluences above the ablation threshold. In these cases, residual stresses increase with increasing fluence or number of laser pulses.

For a special class of short wavelength sources such as the vacuum-ultraviolet Raman laser [6-9] and the 157-nm F$_2$ laser [10-12], precise etch-rate control and smooth morphology are available over a wide fluence range without deleterious effects such as microcracking or bulk coloration. While prospects for shaping optical materials and fabricating photonic components are highly promising, bending and distortion of small fused silica samples has been noted in our laboratory during 157-nm laser ablation. Laser ablation provides rapid heating of thin melt zones and glassy-temperature zones that generally supports stress relaxation. This is followed by residual tensile stress formation during the cooling cycle to ambient temperature [16]. The purpose of this paper is to evaluate residual stress formed during crack-free ablation of fused silica with a 157-nm laser. A ~275-nm thick laser-affected layer was noted, which together with a thin-film-on-substrate model [17] of sample bending, yielded an average tensile stress of ~80 MPa.

ABLATION EXPERIMENTS

A F_2-laser (Lambda Physik model LPF 220i) provided 15-ns pulses at 157-nm wavelength. A CaF_2-lens pair (18-cm and 7-cm focal lengths) demagnified the 22-mm x 7-mm beam to provide uniform exposure of a rectangular mask at ~45 mJ/cm^2 fluence. The mask was imaged by a MgF_2 lens (5-cm focal length) using ~15x demagnification to provide spatially uniform fluence (± 5%) in the range of 1.9 - 4.7 J/cm^2 on target samples. These fluences exceed the 1.0 J/cm^2 ablation threshold of fused silica [11]. The beam line and target chamber was evacuated of air and flushed with argon (500 mTorr) to provide transparency at 157-nm. Details of the optical system have been described previously [11-12].

Fused silica cover slips (Corning 7940, UV grade) of 160-μm thickness were cut with a diamond scriber into 0.25-mm by 10-mm rectangles, a geometry offering pronounced bending when laser ablated. Samples were dipped in HF acid (50%) for 90 s to relieve stresses induced by the scribing process. Samples were cleaned with alcohol solvent and mounted onto an x-y translation stage inside the target chamber in the orientation shown in Fig. 1. The laser beam overfilled the sample width and the sample was scanned vertically at speeds of 94 or 220 μm/s. The laser repetition rate was 10 Hz. Vertical and horizontal surface profiles of a static fused silica sample are shown on the left in Fig. 1 for ablation with 25 pulses at 4.7-J/cm^2 fluence. A moderately uniform excision of 2.2-μm depth is noted. Ablation depths for scanned samples varied from 0.15 to 1.0 μm. Ablation debris of 10 to 20 nm thickness typically accumulated over the scanned surface sample for these scanning speeds. Thicker coatings can affect the sample bending but were not thought to influence the observations in the present case.

Sample surface profiles and shapes were characterized with a WYKO optical surface profiler (white-light vertical-scanning interferometry mode) before and after laser ablation. One end of the sample was clamped to stabilize the interferometric fringes without distorting the sample. A stylus profilometer (Tencor) was used for short-range surface morphological measurements.

The thickness of the laser-affected layer was evaluated by etching bent samples (10's of seconds) in hydrofluoric acid (50%) and noting the bending radius as a function of the chemical etch depth. Chemical etching increased the surface roughness slightly to <10 nm (rms).

RESULTS AND DISCUSSION

All ablated samples bent toward the same direction – opposite to the laser beam direction – indicating the development of tensile stresses immediately below the ablation

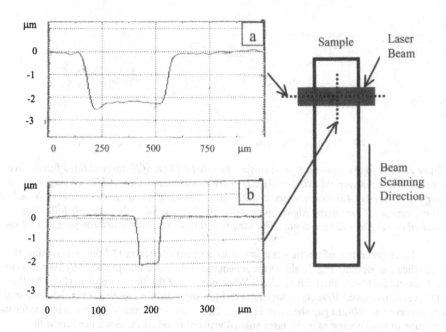

Figure 1. *Horizontal (a) and vertical (b) ablation beam profiles (Tencor stylus profilometer) for 157-nm ablation of fused silica (left) ablated with 25 pulses at 4.7 J/cm². Schematic (right) showing the orientation of the laser-ablation beam relative to scanning fused silica sample. See text for further details.*

surface. Material compaction is anticipated in this layer. Compaction of fused silica is observed when below-ablation fluence irradiates surfaces. Figure 2 (left) shows the convex back surface of an ablated sample with a center deflection of 180 nm across a 4.7-mm span. Surprisingly, a center deflection of 195 ±15 nm was noted for all irradiated samples irrespective of the fluence (1.9 - 4.7 J/cm²), the scanning speed (94 or 220 μm/s), or the ablation depth (0.15 – 1.0 μm) applied to the samples. This consistency suggests the formation of a fixed thickness of laser-affected material. Each additional laser ablation pulse must therefore remove and generate an equal thickness of tensile material in the laser-affected layer.

Single-pulse ablation rates (depth D) for fused silica follow [11] a logarithmic fluence dependence characteristic of strongly absorbing polymers and dielectrics:

$$D = (1 / \alpha) \ln (F / F_{th}) \qquad (1)$$

Here, $\alpha = 1.7 \times 10^5$ cm^{-1} is the effective absorption coefficient, F is the laser ablation fluence, and $F_{th} = 1.0$ J/cm² is the ablation fluence threshold. This relation suggests that a fixed fluence, F_{th}, will penetrate to the ablation-melt interface (at depth D) irrespective of the incident laser fluence. Larger fluence values are balanced by removal of thicker layers of material.

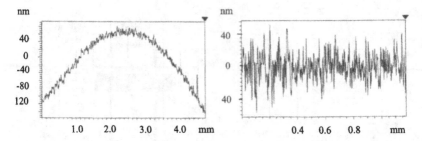

Figure 2. *Left figure shows a typical surface profile (WYKO) of F_2-laser ablated fused silica recorded along the vertical axis (see Fig. 1) of the non-radiated side. A 500-nm deep surface layer was excised at 4.0 J/cm^2 fluence and 220-$\mu m/s$ scanning speed. The center is deflected by 180-nm over a 4.7-mm span. Right figure shows the front surface of this sample following chemical etching to ~300 nm depth. The sample is flat with a rms surface roughness of 20 nm.*

Laser generation of tensile stress layers are also manifested in 157-nm laser cutting of fused silica cantilevers. Fig. 3a shows the schematic outline of a 200-μm wide by 10-mm long slot laser-cut in160-μm thick fused silica. When one end of the slot was broken (Fig. 3b), the cantilever arms moved 10% closer together (180-μm separation) as shown in Fig. 3d compared with the original 200-μm gap shown in Fig. 3c. This inward direction is further evidence for the formation of tensile stresses in the laser-ablated walls of fused silica. Such bending with ablation-induced stresses limits the precision available in laser machining of 3-dimensional silica microstructures.

The material compaction and concomitant tensile stress generation as evidenced in Figures 2 and 3 must occur in thin layers. Thermal diffusion suggests a ~100-μm thick heating zone in fused silica during the 15-ns laser-heating pulse while the effective-absorption coefficient (Equation 1) provides a 59-nm value for the optical penetration depth (during laser ablation). To better assess the laser-affected layer thickness, strained samples were thinned by wet chemical etching until stress relief was noted. Figure 2 (right) shows a sample flattened after etching 300-nm of the sample surface. A slight increase in surface roughness to 20 nm (rms) is noted. The radius of curvature was doubled (deflection halved) when only a 100-nm thick surface layer was chemically removed. These values only slightly exceed the thermal transport and optical penetration depth associated with 157-nm laser ablation, and is much smaller than the 160-μm thickness of the fused silica sample. More precise chemical etching data is one future goal of this group to determine tensile stress relief below the ablated surface.

A thin film having uniform stress, σ, will deflect a long thin substrate to radius, r, according to [17]

$$\sigma = Ed^2 / 6rt. \tag{2}$$

Here, E is Young's modulus for the substrate, d is the substrate thickness, and t is the thickness of the film. Strain associated with tensile stress in the film yields a concave shape on the thin film side, similar to the deflection caused by tensile stresses in the laser-affected layer of the present fused silica strips. If we assume the stress profile in the laser-affect layers of fused silica is

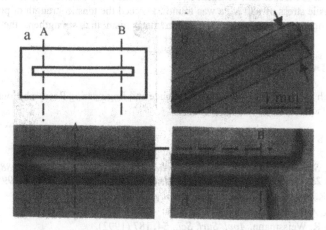

Figure 3. *Schematic and photos of fused silica with a lase- cut trough: (a) drawing of sample with trough before breaking along section B; (b) photo of the sample showing the broken end near B; (c) magnification of 100-μm trough at section A; (d) tensile stresses on laser cut surface is revealed by reduced width (180-μm) of trough at section B after breaking. The optical microscope was imaged on the top surface of the sample.*

uniform across a 275-nm thickness, then the tabulated value of $E = 72$ GPa [18] yields an average residual stress of 80 MPa directly below the laser-ablated surface. Larger stresses are anticipated at the ablated surface, yielding stresses well above the 50 MPa tensile strength of bulk-fused silica [18, 19].

Such a large tensile stress may be stable if 157-nm laser ablation strengthens the near-surface layer (200-300 nm) of fused silica. Tensile strengths up to 6 GPa have been reported [22, 23] in undamaged fused-silica fibers. However, for the bulk samples studied here, much lower strengths are anticipated. Further, optical polishing is widely known to induce surface defect states and stresses that weaken the surface. Laser-damage studies of glass windows also demonstrate the importance of surface treatments and underlying stresses on the value of the damage threshold – typically noted as the onset of microcrack formation. For example, the 325-nm laser-damage threshold for thermally treated glasses (fused silica and BK-7) is 2–4 factors less than untreated samples [20], while cleaved fused silica generally offers comparably higher damage threshold [21]. Such surface treatment does not appear to play a significant role in 157-nm laser ablation of fused silica [10, 11]. Surface damage and microcracks are rarely observed with this source suggesting a different mechanism that possibly strengthens the underlying ablated surface. Laser treatments that strengthen optical glasses are highly desirable and deserve further study.

CONCLUSIONS

F_2-laser ablation of fused silica generates localized tensile stresses in a thin ~275-nm surface layer that is independent of fluence (1.9- 4.7 J/cm^2) and ablation depth (150-1000 nm). An inferred tensile stress of ~80 MPa was found to exceed the tensile strength of polished fused silica, suggesting the a possible laser ablation mechanism that strengthens the fused silica surface.

ACKNOWLEDGEMENTS

Research was supported by National Science and Engineering Research Council (Canada) and Photonics Research Ontario.

REFERENCES

1. *Thin Films: Stresses and Mechanical Properties I to VII, MRS Proc. 130 (1988); 188(1990); 239(1991); 308(1993); 356(1994); 436(1996); 505(1997); 646(1998).*

2. B. Braren, R. Srinivasan, *J. Vac. Sci. Technol.*, **B6**, 537 (1988).

3. J. Ihlemann, *Appl. Surf. Sci.*, **54**, 193 (1992).

4. C. Buerhop R. Weissmann, *Appl. Surf. Sci.*, **54**, 187 (1992).

5. P.E. Dyer, R.J. Farley, R. Giedl D.M. Karnakios, *Appl. Surf. Sci.*, **96-98**, 537 (1996).

6. J. Zang, K. Sugioka, S. Wada, H. Tashiro, K. Toyoda, *Jpn. J. Appl. Phys.* 35, L1422 (1996).

7. Y. Tkigawa, K. Kurosawa, W. Sasaki, M. Okuda, K. Yoshida, E. Fujiwara, Y. Kato, Y. Inoue, *J. Non-Crystal. Solids*, **125**, 107 (1990).

8. K. Sugioka, S. Wada, Y. Ohnuma, A. Nakamura, H. Tashiro, K. Toyoda, *Appl. Surf. Sci.*, **96-98**, 347 (1996).

9. K. Sugioka, S. Wada, A. Tsunemi, T. Sakai, H. Takai, H. Moriwaki, A. Nakamura, H. Tashiro, K. Toyoda, *Jpn. J. Appl. Phys.* **32**, 6185 (1993).

10. P.R. Herman, B. Chen, D.J. Moore, M. Canaga-Retnam, *MRS Proc.* **236**, 53 (1992).

11. P.R. Herman, K. Beckley, B. Jackson, K. Kurosawa, D. Moore, T. Yamanishi, J. H. Yang, *SPIE Proc.* **2992**, 86 (1997).

12. P. R. Herman, R.S. Marjoribanks, A. Oettl, K. Chen, I. Konovalov, S. Ness, *Appl. Surf. Sci.*, **154-155**, 577 (2000).

13. N.F. Borrelli, C. Smith, D.C. Allan, T.P. Seward, *J. Opt. Soc. Am.*, **B14**, 1606 (1997).

14. E. R. Cochran, C. Ai, *Appl. Optics*, **31**, 6702 (1992).

15. F. Dahmani, A.W. Schmid, J.C. Lambropoulos, S. Burns, *Appl. Optics*, **37**, 7772 (1998).

16. F. Guignard, M.L. Autric, V. Baudinaud, *SPIE Proc.*, **3343**, 534 (1998).

17. R.W. Hoffman, in 'Physics of Thin Films', ed. G Hassand and R.E. Thun, (Academic Press Inc., New York, 1966) 3, pp.211- 272.

18. W.C. Oliver, G.M. Pharr, *J. Mater. Res.*, 7, 1564 (1992).

19. F. Dahmani, J.C. Lambrpoulos, A.W. Schmid, S. Papernov, S.J. Burns, *J. Mater. Res.*, 14, 597 (1999).

20. O.M. Efimov, L. Glebov, S. Papernov, A.W. Schmid, *SPIE Proc.*, **3578**, 564 (1999).

21. S. Papernov, D. Zaksas, J.F. Anzellotti, D.J. Smith, A.W. Schimid, D.R. Collier, F.A. Carbone, *SPIE Proc.*, **3244**, 434 (1998).

22. S. Sakaguchi T. Kimura, *J. Am.Ceram.Soc.*, **64**, 259 (1981).

23. H.C. Chandan D. Kalish, *J. Am.Ceram. Soc.*, **65**, 171 (1982).

Mat. Res. Soc. Symp. Vol. 617 © 2000 Materials Research Society

Laser Induced Fluorescence Measurement of Plasma Plume During Pulsed Laser Deposition of Diamond-Like Carbon

Yukihiko Yamagata, Yuji Kozai, Fumiaki Mitsugi, Tomoaki Ikegami, Kenji Ebihara, Ajay Sharma[1], Robert M. Mayo[2] and Jagdish Narayan[1]
Department of Electrical and Computer Engineering, Kumamoto University,
Kurokami 2-39-1, Kumamoto 860-8555, JAPAN
[1] NSF Center of Advanced Materials and Smart Structures, North Carolina State University,
Raleigh, NC 27695-7916, U.S.A.
[2] Department of Nuclear Engineering, North Carolina State University,
Raleigh, NC 27695-7909, U.S.A.

ABSTRACT

Dynamics of carbon ablation plasma plume during the preparation of diamond-like carbon films by KrF excimer pulsed laser deposition was investigated using laser induced fluorescence (LIF) and optical emission spectroscopy. LIF signal from C_2 molecule (Swan band, $d\,^3\Pi_g - a\,^3\Pi_u$) was detected using a photomultiplier tube and an intensified CCD camera. Temporal evolution and spatial distribution of C_2 molecules in the ablated plume were measured as a function of laser energy density and ablation area. LIF intensity is found to be weaker in the central part of the plume than that at the periphery at incident energy greater than 6 J/cm^2. It is conjectured that some of C_2 molecules are dissociated by collision with energetic species in central part of the ablation plume. Dynamics of ablation plasma plume is strongly dependent on the size of ablated area.

INTRODUCTION

Hydrogen-free diamond-like carbon (DLC) films have great potential for applications in mechanical and optical coatings, electronic devices, and field emitters because of their attractive properties, such as extreme hardness, chemical inertness, very low electrical conductivity, optical transparency over a wide range, and low electron affinity. Pulsed laser deposition (PLD) has many advantages of low temperature processing, low levels of contamination, stoichiometry preservation, and reproducibility of film characteristics compared with other deposition methods. Since the average energy of laser ablated species is much higher than kT and a certain fraction of these species is ionized, it is possible to form metastable phases such as DLC [1-8], high fullerene carbon molecules of C_{60}, C_{70} and C_{84}, carbon nanotubes, and carbon nitride. Consequently, PLD has been regarded as one of the most successful techniques to synthesize novel metastable materials and structures.

The dynamics of laser ablated plasma plume affects the characteristics of deposited thin films. To understand the PLD process and to correlate the properties of the deposited films with that of plasma parameters the ablated plume has been investigated using optical emission spectroscopy (OES) [1-4], the Langmuir probe [4-6], laser-induced fluorescence (LIF) spectroscopy [4,9,10], time of flight mass spectrometry [7] and interferometry [8]. In spite of several investigations, however, the mechanism of the PLD process has not yet been understood, especially in the case of DLC deposition. In order to prepare high-quality thin films and nanoparticles and to improve the potential of PLD, it is desirable to diagnose and establish a correlation between plasma composition and properties of thin films.

In this paper, we describe quantitative study on ablation plasma using LIF spectroscopy and OES. Temporal evolution and spatial distribution of C2 molecule are measured by LIF as a function of ablation laser parameters using fast photography.

EXPERIMENTAL DETAILS

The carbon plasma was created in a stainless steel vacuum chamber pumped to a base pressure of 10^{-6} Torr by irradiating a carbon target (Kohjundokagaku, purity 99.999%) placed at an angle of 45 using a KrF excimer laser (Lambda Physik ComPex205, λ=248nm, pulse duration of ~20ns). The incident laser energy and the ablation area were varied from 300 – 700 mJ and $1.3\times4.1 - 2.3\times4.4$ mm^2, respectively. This corresponds to the laser energy density (E_L) of 2.0 – 10.0 J/cm^2 on the target. A dye laser beam from a dye laser (Lambda Physik SCAN mate 2EC-400) pumped by third harmonic of Nd:YAG laser tuned at 516.52 nm was sent through the carbon plume parallel to the target surface to excite C$_2$ molecules ((0, 0), $a\,^3\Pi_u - d\,^3\Pi_g$). The dye laser beam can be delayed with respect to the ablating excimer laser pulse. The laser induced fluorescence (LIF) at 563.49 nm corresponding to a C ((0,1), $d\,^3\Pi_g - a\,^3\Pi_u$) transition was detected at right angle to both the ablating and the probe dye laser beam. Since $a\,^3\Pi_u$ level is very close to C$_2$ ground state the LIF intensity is proportional to the C$_2$ number density. Fluorescence from the carbon ablation plume was collected by a quartz lens (f=100 mm) and imaged onto an entrance slit of a monochromator/spectrograph (ARC SpectraPro-308i) using an optical fiber bundle. Time-resolved LIF and spontaneous emission from the ablation plume were measured using a photomultiplier (Hamamatsu R928) and a digitizing oscilloscope (Tektronix TDS3034). LIF was measured as a function of distance from the target surface (d) and delay time (τ_d) of the probe laser.

In order to record two-dimensional LIF (2D-LIF) images probe laser beam was shaped like a sheet of 20 mm width and about 0.5 mm thickness using cylindrical lenses. LIF images were taken using a gated and intensified CCD camera (PI ITE/CCD) with gate time of 20 ns and a narrow band pass interference filter (λ=570 nm, $\Delta\lambda$=10 nm). The time-resolved LIF and emission profiles, and 2D-LIF images were accumulated over 20 laser shots to improve the signal to noise ratio.

RESULTS AND DISCUSSION

Figure 2 shows a typical waveform of time-resolved emission at λ=563.5nm corresponding to optical emission from C$_2$ ((0,1), $d\,^3\Pi_g - a\,^3\Pi_u$) recorded at 1 mm from the target surface. The ablation plume was produced at laser energy density of 6 J/cm^2 and S=1.3x4.1 mm^2. Fig also shows KrF laser pulse, probe dye laser pulse at delay of 300 ns with respect to KrF laser pulse. The temporal profile

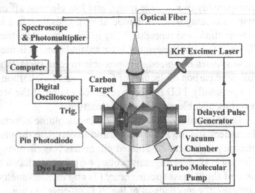

Figure 1. Experimental setup for LIF measurement of ablation plasma plume.

has two peaks. The first broad peak is the sum of spontaneous emission from C_2 molecules and the background emission corresponding to the recombinative radiation that is intense near the target surface. The second peak is the LIF signal from C_2 molecules and several–fold larger than the spontaneous emission.

The LIF intensity changes with τ_d, d, E_L and S. Figure 3 shows LIF intensity at d=1, 5, 10 mm as a function of delay time τ_d. In order to see the effect of ablation area, the laser energy density was kept constant at 6J/cm^2 and the area was changed from 1.3x4.1 to 2.3x4.4 mm^2. In both cases, the LIF intensity (number density of C_2) decreases with increase in delay time. The LIF intensity for

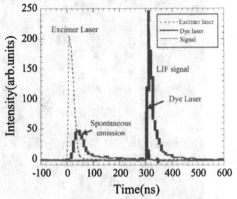

Figure 2. Typical waveform of time-resolved emission at λ=563.5nm.

S=1.3x4.1 mm^2 decreases rapidly with the increase in d, while that for S=2.3x4.4 mm^2 decreases slightly. It suggests that temporal and spatial distribution of C_2 molecules change drastically with the ablation area S.

Figures 4 shows temporal images of spontaneous emission LIF from C_2 molecules. The ablation plumes were produced with the same E_L of 6 J/cm^2 on ablation areas of S=1.3x4.1 mm^2 (Figs. 4(a) and 4(c)) and S=2.3x4.4 mm^2 (Figs. 4(b) and 4(d)), that were used for temporal waveform measurements in Figs. 3(a) and 3(b), respectively. Spontaneous emission is observed

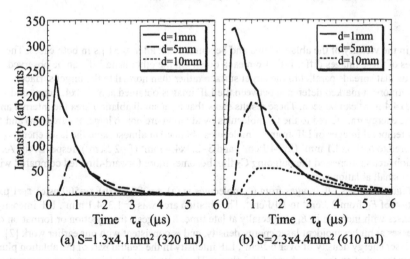

(a) S=1.3x4.1mm^2 (320 mJ)   (b) S=2.3x4.4mm^2 (610 mJ)

Figure 3. LIF intensity change at d=1, 5, 10 mm as a function of delay time τ_d. E_L=6 J/cm^2, (a) S=1.3x4.1 mm^2, (b) S=2.3x4.4 mm^2.

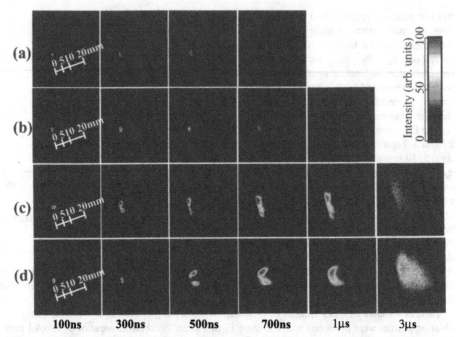

Figure 4. LIF and spontaneous emission images from C_2 molecules at different ablation areas under constant laser fluence of 6 J/cm^2. Spontaneous emission : (a) S=1.3x4.1mm², (b) S=2.3x4.4mm², LIF : (c) S=1.3x4.1mm², (d) S=2.3x4.4mm².

only in the vicinity of the ablation area and becomes weak after $\tau_d > 1$ μs in both cases. The shapes of these images differ little from each other. On the other hand LIF can be observed at $\tau_d \sim 4$ μs that spreads parallel to the target surface rather than normal to the target surface. Furthermore, a marked deference between the LIF images obtained at S=1.3x4.1 mm² and S=2.3x4.4 mm² can be seen. These results show that for a small ablation area the internal and kinetic energy transferred to the C_2 molecules by ablation are not so large. Figures 5(a) and 5(b) show temporal images of LIF from C_2 molecules obtained at almost same the laser energy delivered to S=1.3x4.1 mm² (E_L=4 J/cm²) and S=2.3x4.4 mm² (E=2 J/cm²), respectively. As the ablation area is increased the outgoing C_2 flux becomes more forward-directed compared with that of small ablation area.

Figure 6 shows LIF images from C_2 molecules at (a) τ_d=300 ns, (b) τ_d=500 ns,c) τ_d=1 μs as a function of E_L from 2 J/cm² to 10 J/cm². The ablation area was S=1.3x4.1 mm². LIF intensity decreases with increase in E_L, especially at late time. It suggests that ejection or formation of C_2 decreases at higher incident laser energy density, and is consistent with our earlier work [2]. At high laser energy density of over 6 J/cm², LIF intensity in the central part of the ablation plume is weaker than that on the periphery of the plume. This is attributed to large surface temperature of the target where C_2 molecules are dissociated by collision with energetic species in the central

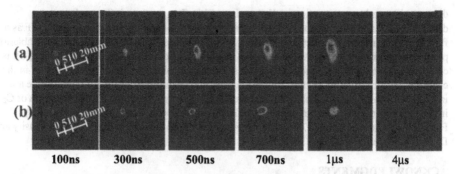

| | 100ns | 300ns | 500ns | 700ns | 1μs | 4μs |

Figure 5. Temporal C_2 LIF images obtained at almost same laser energy delivered to (a) S=1.3x4.1 mm² (4 J/cm²) and (b) S=2.3x4.4 mm² (2 J/cm²).

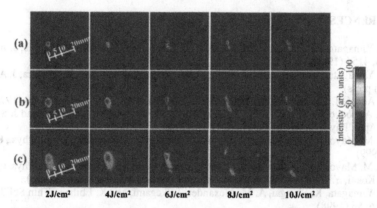

| | 2J/cm² | 4J/cm² | 6J/cm² | 8J/cm² | 10J/cm² |

Figure 6. LIF images from C_2 molecules at (a) τ_d=300 ns, (b) τ_d=500 ns, (c) τ_d=1 μs as a function of E_L.

part of the ablation plume. It is well known that high laser energy density is necessary to synthesize good DLC film with high fraction of sp^3 bond using PLD method. Present LIF results show that it is important to suppress the C_2 formation for high quality DLC preparation, and support our previous works [1,2,5,6]. It is also suggested that plasma plume dynamics is strongly influenced by the size of ablation area in addition to laser energy density. Ablation area is an important parameter in PLD and has to be considered to understand the carbon ablation phenomena.

CONCLUSIONS

Laser induced fluorescence was employed to investigate carbon ablation plasma plume during diamond-like carbon preparation by KrF excimer PLD. Temporal evolution and spatial distribution of C_2 molecules in the ablation plume were measured by two-dimensional LIF as a function of laser energy density and ablation area. LIF intensity in the central part of the ablation plume is weaker than that at the surrounding under high laser energy density, over 6 J/cm^2. It is conjectured that some of C_2 molecules are dissociated by collision with energetic species in the central part of the ablation plume. Number density and spatial distribution of C_2 molecule are influenced by laser energy density and ablation area. To know formation and dissociation of C_2 in the plume, a measurement of density profiles of other species like C and C^+ is necessary. Further measurement using 2D-LIF and investigation of relationship between species density and DLC film properties are being undertaken.

ACKNOWLEDGMENTS

A part of this research was financially supported by the Grant-in-Aid for Scientific Research of JSPS [No.11680489].

REFERENCES

1. Y. Yamagata, A. Sharma, R. M. Mayo, J. W. Newman, J. Narayan, K. Ebihara, J. Appl. Phys., **86**, 4154 (1999).
2. Y. Yamagata, A. Sharma, R. M. Mayo, J. W. Newman, J. Narayan, K. Ebihara, J. Appl. Phys. (in press).
3. S. Aoqui, T. Ikegami, Y. Yamagata, and K. Ebihara, Thin Solid Films **316**, 40 (1998).
4. A. A. Voevodin, S. J. P. Laube, S. D. Walck, J. S. Solomon, M. S. Donley, and J. S. Zabinsky, J. Appl. Phys. **78**, 4123 (1995).
5. R. M. Mayo, J. W. Newman, A. Sharma, Y. Yamagata, J. Narayan, J. Appl. Phys., **86**, 2865 (1999).
6. R. M. Mayo, J. W. Newman, A. Sharma, Y. Yamagata, J. Narayan, J. Appl. Phys. (in press).
7. F. Kokai, Y. Koga, and R. B. Heimann, Appl. Surface Sci. **96-98**, 261 (1996).
8. Y. Yamagata, K. Shingai, A. M. Alexander, T. Ikegami, and K. Ebihara, Thin Solid Films **316**, 56 (1998).
9. Y. Nakata, H. Kaibara, T. Okada, and M. Maeda, J. Appl. Phys. **80**, 2558 (1996).
10. D.B. Geohegan and A.A. Puretzky, Appl. Phys. Lett. **67**, 197 (1995)

Mat. Res. Soc. Symp. Vol. 617 © 2000 Materials Research Society

STRUCTURAL CHARACTERIZATION OF LASER LIFT-OFF GaN

ERIC A. STACH,* M. KELSCH,*,# W.S. WONG,†,∫ E.C. NELSON,* T. SANDS† AND N.W. CHEUNG§

* National Center for Electron Microscopy, Materials Science Division, Lawrence Berkeley National Laboratory, Berkeley, CA 94720: email: EAStach@LBL.gov ; http://ncem.lbl.gov

† Department of Materials Science and Engineering, University of California, Berkeley, CA 94720;

§ Department of Electrical Engineering and Computer Science, University of California, Berkeley 94720.

\# On leave from the Max Plank Institute für Metallforschung, Stuttgart, Germany.

∫ Present address: Xerox Palo Alto Research Center, Palo Alto, CA 94304.

ABSTRACT

Laser lift-off and bonding has been demonstrated as a viable route for the integration of III-nitride opto-electronics with mainstream device technology. A critical remaining question is the structural and chemical quality of the layers following lift-off. In this paper, we present detailed structural and chemical characterization of both the epitaxial layer and the substrate using standard transmission electron microscopy techniques. Conventional diffraction contrast and high resolution electron microscopy indicate that the structural alteration of the material is limited to approximately the first 50 nm. Energy dispersive electron spectroscopy line profiles show that intermixing is also confined to similar thicknesses. These results indicate that laser lift-off of even thin layers is likely to result in materials suitable for device integration. Additionally, because the damage to the sapphire substrate is minimal, it should be possible to polish and re-use these substrates for subsequent heteroepitaxial growths, resulting in significant economic benefits.

INTRODUCTION

III-nitride semiconductor alloys are promising materials for opto-electronic devices in the ultraviolet to blue/green spectrum. This is because the III-nitrides form a continuous alloy system with direct band gaps over the range of 1.9 eV (InN) to 3.4 eV (GaN) to 6.2 eV (AlN). This has resulted in the successful creation of blue and green laser diodes, as well as the full color spectrum of light emitting diodes.[1,2] However, because of the low decomposition temperature of GaN (on the order of 900 °C), significant problems remain in the growth of materials of high crystal quality. This is because this low decomposition temperature makes bulk crystal growth difficult using standard methods. Additionally, the dissociation of nitrogen from typical carrier gases used in metallorganic chemical vapor deposition (MOCVD) requires high temperatures that are often incompatible with growth on conventional substrates. As a result, the majority of III-nitride devices are grown heteroepitaxially onto either sapphire (single crystal (0001) Al_2O_3), or less frequently, SiC. These two materials provide a hexagonal template for the growth of wurtzite GaN, and can easily withstand high crystal growth temperatures. However, both sapphire and SiC have electrical and thermal conductivity constraints which may limit the functionality of III-nitride devices grown on these materials.

Recently, Kelly et al.[3] and Wong et al.[4,5] have demonstrated a method of epilayer lift-off and bonding that permits direct integration of III-nitride devices with most substrate materials. This method takes advantage of the different band gaps of GaN and sapphire. A pulsed excimer KrF laser at 5 eV (λ = 248 nm) is used to thermally decompose the heteroepitaxial interface between the two materials. At this wavelength, the sapphire substrate is transparent, but this wavelength is also well above the absorption edge of GaN. With sufficient laser fluence (> 400 mJ/cm^2) this absorption results in heating local to the GaN / sapphire interface which causes decomposition of the GaN into metallic Ga and N_2 gas. Slight warming of the material above the Ga melting point thereafter releases the GaN layer from the substrate. Prior work has characterized the bulk properties of the resulting materials using scanning electron microscopy,[5] x-ray diffraction,[5] channeling Rutherford backscattering spectroscopy[6] and photoluminescence.[6] Additional experiments have shown that device layers can function following lift-off and bonding.[5,7] In this paper, we present detailed electron microscopy characterization of both the laser lift-off (LLO) GaN layer and the remaining sapphire substrate. We find that the damage to both the epilayer and substrate is quite minimal, with both structural alteration and chemical intermixing confined to approximately the first 50 to 100 nm of the epilayer and the substrate.

EXPERIMENTAL

The GaN layers were grown heteroepitaxially on (0001) oriented sapphire to a thickness of 12.5 μm using hydride vapor phase epitaxy (HVPE). No growth buffer layer was used. Prior to laser irradiation the substrates were polished to ≈ 3 μm roughness using diamond paste to reduce the scattering of the laser light. The samples were irradiated from the back side of the heterostructure with a single pulse of the KrF laser at a fluence of 600 mJ/cm^2. The heterostructures were then warmed to 40 °C on a hot plate to melt the decomposed Ga interfacial layer and complete the lift-off process.

Cross sectional TEM sample preparation of the thin GaN layers proved difficult due to the different ion thinning rates of GaN, sapphire and the glue used in the preparation process. In order to obtain electron transparent regions of the LLO GaN layers at the location of the prior heterointerface, a modification of an existing TEM sample preparation method was used. A total of seven LLO GaN layers were glued together in succession in a miniature vice (i.e. two layers were glued together first, followed by a third, then a fourth, etc.). This resulted in very thin glue layers – on the order of 0.1 μm to 0.25 μm. The resulting GaN sandwich was then glued between Si (001) substrate material for support. TEM preparation thereafter followed the method of Bravman and Sinclair.[8] Final ion milling was performed using a Technorg Linda low angle, low voltage ion mill. Initial thinning was done at 10 kV and 5° incidence until perforation, followed by a polish at 500 eV and 5° for a half hour. High resolution electron microscopy was performed using the NCEM Atomic Resolution Microscope at 800 kV and analytical electron microscopy was performed using a Philips CM200 field emission microscope equipped with a Kevek atmospheric thin-window energy dispersive spectrometer and the Emispec control and analysis software. Conventional diffraction contrast microscopy was performed using a JEOL 200CX microscope at 200 kV, as well as the ARM at 800 kV.

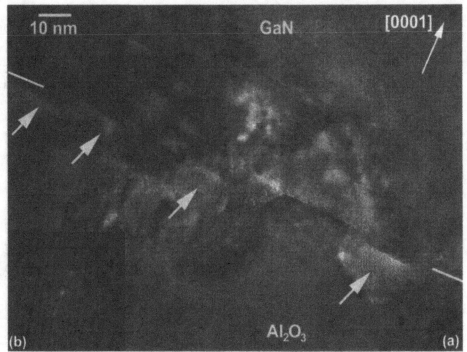

Figure 1 – (a) Large area HREM image of the HVPE GaN / sapphire heterostructure prior to laser lift-off. Arrows point to amorphous regions at the interface. (b) Computed diffractogram of the interfacial region showing the orientation relationship between the two materials.

RESULTS AND DISCUSSION

Figure 1 shows a large area high resolution micrograph (HREM) of the GaN / sapphire heterostructure prior to laser lift-off. Inset is a computed fast Fourier transform (FFT) diffractogram taken from the interfacial region. This diffractogram indicates that the orientation relationship between the substrate and heteroepitaxial layer is $(0001)_{GaN} // (0001)_{Al_2O_3}$ and $[01\bar{1}2]_{GaN} // [01\bar{1}2]_{Al_2O_3}$. Although this orientation relationship has been observed,[9,10] it is much more common to observe $(0001)_{GaN} // (0001)_{Al_2O_3}$ and $[01\bar{1}0]_{GaN} // [01\bar{1}2]_{Al_2O_3}$ in heteroepitaxial GaN / sapphire.[11] Visible at the interface between the two layers are pockets of amorphous material; this amorphous material is a true feature of the heterostructure and not a TEM preparation related artifact. Additionally, numerous stacking faults are observed within the first 30 to 40 nm of the layer. These features, along with the high dislocation densities present in the layer ($\rho \approx 10^{10}$ cm^{-2}, not visible in this image), act to accommodate the heteroepitaxial strain.

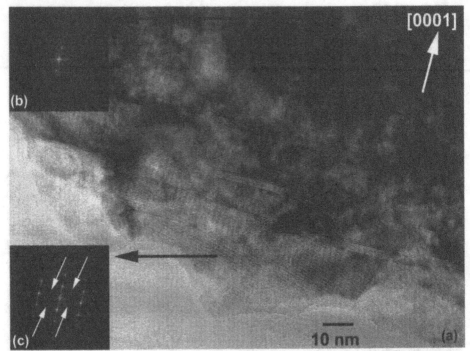

Figure 2 – (a) Large area HREM image of the GaN layer following LLO. (b) Diffractogram from region away from LLO interface. (c) A typical diffractogram from the LLO region. Defect spots are arrowed.

Figure 2 shows a similar large area HREM micrograph of the GaN layer following laser lift-off. This image was taken from the thinnest area of the LLO GaN TEM sample that still had specimen preparation glue remaining (the amorphous feature which lines the bottom of the sample throughout the image). This indicates that we are in fact imaging the sample at the LLO surface. Again, stacking faults are visible in the image within the first 40 to 50 nm of the new surface. This shows that lift-off occurred *at* the site of the GaN / sapphire interface, and not within either the GaN or sapphire bulk. The inset diffractogram (Figure 2.b) from a region away from the LLO surface shows that the bulk of the GaN layer has a structural perfection equivalent to that observed in the sample prior to lift-off (See Figure 1.b for comparison). However, at the newly created surface, the diffractograms indicate that although the majority of the material is wurtzite GaN (the strong reflections in Figure 2.c), there is significant formation of structural defects consistent with twinning. (Due to a lack of resolution in the computed FFT's the exact crystallography of these defects could not be determined.) It is apparent from this image that the structural alteration to the layer is minimal, and confined to the first 50 or so nanometers of the LLO GaN layer.

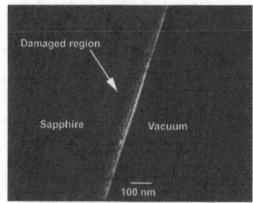

Figure 3 presents a diffraction contrast dark-field electron micrograph of the sapphire substrate following laser lift-off. The mottled contrast at the uppermost region of the sample nearest the newly created surface is consistent with the presence of damage. Unfortunately, the sample preparation process did not produce a region thin enough for high resolution electron microscopy. As a result the exact nature of the damage (amorphization, etc.) is not certain. Further investigation is in progress. Again, though, it is apparent that the structural damage is confined to a region very near ($\approx$ 100 nm) the new surface.

Figure 3 – Dark field image of the sapphire substrate following laser lift-off.

In Figure 4 we present the results of the chemical characterization of both layers. The energy dispersive spectroscopy (EDS) analyses were performed on the Phillips CM200 FEG-TEM, and the spectra were obtained by scanning a 1.6 nm diameter electron probe along the [0001] growth direction of the samples. Each scan consisted of 100 points, with a 10 second dwell time at each point. In Figure 4.a, the EDS analyses of the as-grown GaN / sapphire structure indicates that the initial interface is very abrupt, with the slight spread of the data across the interface a result of both finite probe effects and x-ray fluorescence. In Figure 4.b through 4.d the EDS profiles of both the GaN LLO layer (4.b and 4.c) and the sapphire substrate (4.d) are presented. The difference in signal counts between the aluminum and gallium and that of oxygen and nitrogen is due to differences in detector efficiency for the light elements. In each of these spectra it is apparent that intermixing is not significant, and that it is confined to the first $\approx$ 50 nm.

CONCLUSIONS

Conventional, high resolution and analytical electron microscopy have been used to characterize laser lift-off HVPE GaN layers and the remaining sapphire substrate. It is observed that structural damage and chemical intermixing resulting from laser processing is minimal and that is confined to approximately the first $\approx$ 50 nm of the resulting materials. These results indicate that LLO of III-nitride opto-electronic devices represents a viable route for materials integration.

ACKNOWLEDGEMENTS

The samples were provided by James Ren of American Xtal Technology, Fremont CA. The work at NCEM was supported by the Director, Office of Energy Research, Office of Basic Energy Sciences, Materials Science Division of the U.S. Department of Energy under Contract No. DE-AC03-76SF000098. This work was also supported in part by the University of California MICRO Program (Award #98-133). The authors would like to thank C. Kisielowski and C.J. Echer at NCEM and K.T. Moore at Johns Hopkins Univ. for helpful commentary.

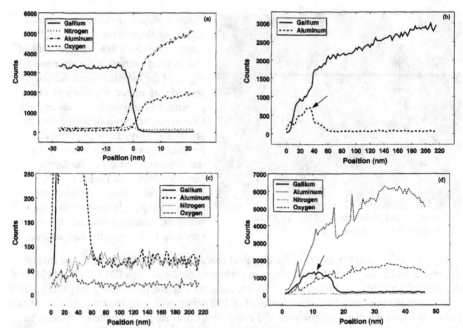

Figure 4 – (a) Energy dispersive spectroscopy line profiles of the as grown heterostructure. (b) EDS from the LLO GaN layer showing distribution of gallium and aluminum. (c) EDS from LLO GaN layer showing all elements. (d) EDS from the sapphire substrate showing all elements.

1 S. Nakamura, et al. MRS Int. J. Nitride Semi. Res., 4S1, 1999.

2 For a review, see S.P Denbaars, Proc. IEEE; 85, 1740 1997.

3 M.K. Kelly, O. Ambacher, R. Dimitrov, R. Handschuh, M. Stutzmann; phys. stat. sol. a. 159, R3-4, 1997; Kelly, et al.; Jap. J. Appl. Phys., Part 2.**38**, (3A) p.L217, 1997.

4 W.S. Wong, T. Sands, and N.W. Cheung, Appl. Phys. Lett. 72, 599 (1998).

5 W.S. Wong, T. Sands, N.W. Cheung, M. Kneissl, D.P. Bour, P. Mei, L.T. Romano and N.M. Johnson, Appl. Phys. Lett. 75, 1360 (1999).

6 W.S. Wong, Y. Cho, E.R. Weber, T. Sands, K.M. Yu, J. Krüger, A.B. Wengrow and N.W. Cheung, Appl. Phys. Lett. 75, 1887 (1999).

7 W.S. Wong, A.B. Wengrow, Y. Cho, A. Salleo, N.J. Quitoriano, N.W. Cheung, and T. Sands, J. Electron. Mater. 28,1409 (1999).

8 J.C. Bravman and R. Sinclair, J. Elect. Mic. Tech. 1, 53, 1984.

9 J.A. Wolk, K.M. Yu, E.D. Bourret-Courchesne and E. Johnson, Appl. Phys. Lett. 70, 2268 (1997).

10 H. Selke, S. Einfeldt, U. Birkle, D. Hommel and P.L. Ryder, in Microscopy of Semiconducting Materials, 10th vol. Oxford (1997).

11 V. Potin, P. Vermaut, R. Ruterana and G. Nouet, J. Elect. Mat. 27, 266 (1998).

Mat. Res. Soc. Symp. Vol. 617 © 2000 Materials Research Society

HIGHLY IONIZED CARBON PLASMA GENERATION BY DUAL-LASER ABLATION FOR DIAMOND-LIKE CARBON FILM GROWTH

S. WITANACHCHI, A. M. MIYAWA, AND P. MUKHERJEE
Laboratory for Advanced Materials Science and Technology (LAMSAT), Department of
Physics, University of South Florida, Tampa, FL 33620

ABSTRACT

Carbon plasmas produced by excimer laser ablation show a low ionization yield of about
8-10%. The coupling of a second CO_2 laser pulse into the plasma in the dual-laser ablation
process significantly increases the plasma temperature and the ionization. The resulting rapid
expansion of the plasma gives rise to high ion kinetic energies and broader ion expansion
profiles [1]. Optical emission spectroscopy and an ion probe have been used to investigate the
dynamics of the carbon plasma. Single and dual-laser ablated carbon plumes have been
deposited on DC-biased silicon substrates to form amorphous carbon films. The diamond-like
behavior of these films was studied by Raman spectroscopy. The Raman spectra were
deconvolved to gauge the effect of the density and the energy of ions on the formation of
diamond-like sp^3 -bonded carbon (DLC) films. The advantage offered by the dual-laser process
for the growth of DLC films is discussed.

INTRODUCTION

The unique properties offered by diamond, which include extreme hardness, high
electrical resistivity, high dielectric strength, and high thermal conductivity, have motivated the
deposition of diamond thin films for optical and electronic applications. Plasma enhanced
chemical vapor deposition (CVD) is the most widely used technique for diamond film growth.
Ion beam deposition, biased sputter deposition, and laser ablation are some of the alternative
techniques used for the growth of diamond films[2-5]. These techniques can yield both
crystalline and amorphous structures in which the properties of the films are determined by the
nature of the atomic bonding. The three-fold coordinated (sp^2) bonding leads to graphitic
behavior while four-fold coordinated (sp^3) bonding leads to diamond-like characteristics. The
amorphous network of carbon with approximately tetrahedral local atomic configuration exhibits
diamond-like properties and is thus technologically important. The amorphous diamond-like
carbon (DLC) films deposited by various techniques contain an atomically mixed structure of
sp^2- and sp^3 -bonded carbon. It has been shown that the metastable sp^3 hybridization of carbon is
facilitated by the thermal agitation and shock wave accompanying ion bombardment on the
depositing substrate[6]. Stabilization of the metastable atomic arrangement results from the high
quenching rates associated with the rapid collapse of the thermal spikes produced by the ions[7].
Therefore, atomic agitation and rapid quenching caused by energetic species during deposition is
important for the growth of DLC films. Carbon ionic energies in the range of 80-100 eV have
been shown to produce DLC films with a high percentage of sp^3 hybridization[8]. However,
much higher ionic energies cause displacement damage and surface heating which transforms the
diamond-like sp^3 bonds to graphitic sp^2 bonds[9].

The energetic advantage of pulsed laser ablated carbon plasmas have been used by
several researchers to fabricate DLC films[10,11]. Since the species energy in pulsed laser

ablation can be controlled in the range of 1-100 eV by the laser fluence at the target, this technique is ideally suited for the growth of DLC films. However, the percentage of ionization of a typical excimer laser ablated plume is between 4-10%. The dual-laser ablation process developed in our laboratory has been shown to produce plasmas with a high percentage of ionization[12]. In addition, the high temperature of the generated plasmas leads to rapid expansion. This imparts high kinetic energy to the ions and at the same time broadens the expansion profile. In this investigation, we have used optical emission spectroscopy and a time-of-flight ion probe technique to comparatively study the dynamic expansion of the ions produced in the two laser processes. DLC films have been deposited on biased silicon substrates by both single and dual-laser processes to gauge the effect of the enhanced plume ionization on the sp^3 hybridization of the DLC films. The Raman spectra of the films have been deconvolved into two distinct vibronic modes corresponding to sp^2- and sp^3 -bonded carbon. Based on this information, the advantages of dual-laser deposition for the growth of DLC films have been demonstrated.

EXPERIMENTAL PROCEDURE

Carbon plasma plumes have been generated in vacuum by ablating a high-density carbon target by the conventional excimer laser process and by dual-laser ablation. In the dual-laser ablation process, the excimer laser pulses of wavelength 248nm (KrF) and CO_2 laser pulses of wavelength 10.6µm are spatially overlapped on the target with an inter-pulse temporal delay of about 50 ns. The details of this process are published elsewhere[13,14]. The experimental set up for film growth is shown in Figure 1. The excimer laser fluence at the target was about 2 J/cm^2 while the fluence of the CO_2 laser was about 5 J/cm^2. Under optimum conditions the ablation of material is caused by only the excimer laser pulse whereas the CO_2 laser pulse is absorbed into the material plasma leading to high plasma temperatures. The optical emission of the plasma plume was imaged 1:1 onto an optical fiber which was placed in the focal plane to collect emission from a point on the axis of the

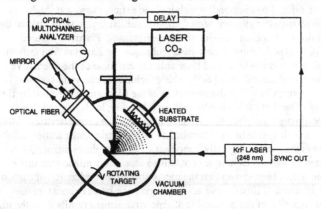

Figure 1: The dual-laser ablation system with provisions for optical emission spectroscopy of the plasma plume

plume. The fiber was connected to the spectrometer of an optical multi-channel analyzer (OMA) system, which recorded the spectra of the excited ionic and atomic species in the plume. An ion probe placed about 6 cm from the target also analyzed the laser-generated plasma. The probe was an insulated copper wire 1mm in diameter, with only its tip exposed to the plume. It was biased at a voltage of −10V. The time-of-flight ion profiles were obtained by using a 500 MHz oscilloscope. The time-of-flight ion data were recorded while translating the probe in 0.5cm steps in both horizontal and vertical directions with respect to the plume. Films of carbon were deposited on solvent-cleaned silicon substrates at room temperature. The bonding of the carbon atoms in the films was investigated by Raman spectroscopy using as excitation wavelength of 514 nm.

RESULTS AND DISCUSSION

The optical emission collected from a point 5 cm from the target on the axis of the expanding plasma plume, for both single and dual-laser ablation is shown in Fig. 2. Most of the line emissions in single laser ablated plume are from excited atoms while high intensity emission from singly and doubly ionized carbon ions dominate the spectrum of the dual-laser ablated plume. In addition to the observed enhancement in the plume ionic content, increases in carbon ion expansion velocities were also observed for dual-laser ablated plumes. The time-of-flight ion probe intensity profiles at two points located 6cm from the target are shown in Fig. 3. In

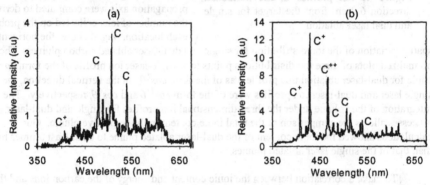

Figure 2: Optical emission from the laser-generated carbon plume 5.0cm from the target for (a) single laser ablation and (b) dual laser ablation.

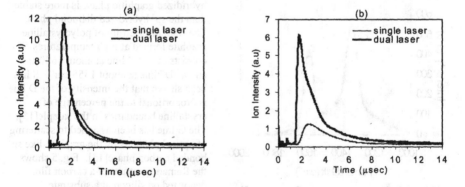

Figure 3: Time of flight ion probe signals 6.0cm from the target (a) on plume axis and (b) 2.5cm from the axis of the plume.

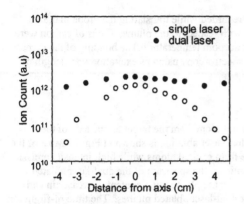

Figure 4: Ion profiles in the horizontal direction 6.0 cm from the target for single and dual laser ablation

addition to enhancement in the ionic content of the plume, the kinetic energy of the carbon ions have undergone a significant increase in dual-laser ablation. Based on the ion profiles presented in Fig.3, the 5eV peak energy of the on axis carbon ions in the single laser ablated plume has been increased to 25 eV by the dual-laser process. Similarly, the ions propagating about 23 degrees off-axis have kinetic energies of 3 eV and 22 eV for single and dual laser ablated plumes, respectively. The area under the time-of-flight ion profiles obtained by scanning the horizontal and vertical directions perpendicular to the plume propagation axis were computed to derive the number of ions collected on the probe at each location. Fig. 4 shows the horizontal spatial variation of the ion distribution for single and dual-laser ablated carbon plumes. The normalized plots of these ion distributions points to a single-laser ion profile of the form $cos^9\theta$ while for dual-laser ablation this profile was of the form $cos^3\theta$. In the vertical direction, the single laser and dual-laser ion profiles were of the form $cos^{14}\theta$ and $cos^6\theta$, respectively. The integration of the volume under the three-dimensional ion profiles for single and dual-laser processes allows the comparison of the total ionic content in the respective plume. Our calculations showed the ionic content of the dual-laser ablated plume to be at least 5 times higher than that of the single laser ablated plumes.

To make a correlation between the ionic content and energy of the carbon ions and the quality of the DLC films, the films deposited on silicon substrates were analyzed by micro-

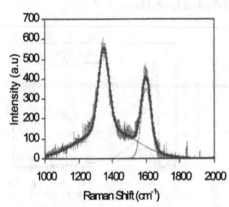

Figure 5: Raman spectrum of a carbon film deposited on a Si substrate at 500°C.

Raman spectroscopy. It is well known that at high temperatures the sp^2 - hybridized graphitic phase is more stable than the sp^3 hybridized diamond phase. The Raman spectrum of polycrystalline graphite formed at high temperatures consists of a "G" line at about 1600 cm^{-1} and a "D" line at about 1355 cm^{-1}. It has been shown that the intensity of the D line is proportional to the percentage of crystalline boundaries in the sample[15]. The G line has been assigned to scattering of optical phonons by the graphite-like sp^2 bonded carbon phase[15]. Fig. 5 shows the Raman spectrum of a carbon film deposited on silicon at a substrate temperature of 500°C. Based on the above

analysis, the carbon in this film is 100% sp^2-hybridized.

The Raman spectra of the amorphous carbon films deposited on silicon substrates at room temperature are shown in Fig. 6. Both spectra contain a main peak around 1530 cm^{-1} and a broad shoulder at around 1350 cm^{-1}. As indicated in the Fig. 6, both spectra have been deconvolved into two components. In several experimental and theoretical investigations of amorphous DLC films, the high frequency peak (G line) has been assigned to sp^2-bonded carbon clusters of fused sixfold rings while the low frequency portion has been interpreted in terms of scattering by sp^3-bonded interconnecting carbon and possible contributions from the graphite-like D peak[14]. We have deconvolved the Raman spectra into three gaussian components, a component peaked at 1600 cm^{-1}, a component peaked around 1350 cm^{-1} and a component peaked around 1530 cm^{-1}. The gaussian function used was of the form

$$I = I_0 \exp\left(\frac{-(\omega - \omega_0)^2}{FW^2}\right) \qquad (1)$$

where, I_0 is the intensity at the peak frequency ω_0, while I is the corresponding intensity at the frequency ω, FW is a constant that corresponds to the full width at half maximum of the distribution[16]. Based on this analysis, the peak centered around 1600 cm^{-1} mainly represents the sp^2 hybridized carbon bonds corresponding to graphite structure (Phase A), similar to the Raman spectrum of the pure graphite film (Fig. 5). The broad low frequency portion (Phase B), which was formed by combining the two components peaked around 1350 cm^{-1} and 1530 cm^{-1}, corresponds to the combination of sp^2- and sp^3-bonded carbon. Therefore, the ratio of the area under the curves corresponding to the two phases can be used to compare the sp^2- and sp^3-bonded carbon content of the deposited films[15]. For the single-laser deposited film, the ratio (Phase B/Phase A) is about 2.4 whereas the ratio for the dual–laser deposited film is about 7.25. This analysis clearly indicates the presence of a high percentage of sp^3-bonded carbon in the dual-laser deposited films.

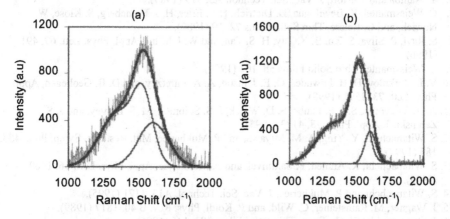

Figure 6: Raman spectra of amorphous carbon films deposited on Si substrates at room temperature by (a) single laser ablation, and (b) dual laser ablation. The solid lines indicate the two deconvoluted phases using gaussian fitting.

CONCLUSION

In this investigation we have demonstrated that the density of carbon ions present in the dual-laser ablated plumes is more than five times higher than the ion density in single laser ablated plumes. In addition, the peak kinetic energy of the ions has been increased by a factor of six in the dual-laser ablation process. Enhanced ion flux and the ionic energy at the depositing substrate is known to promote the formation of sp^3-hybridized carbon by creating localized thermal spikes. Deposition at room temperature rapidly quenches the thermal spikes preventing the transformation of diamond-like sp^3-bonded carbon into graphite-like sp^2-bonded carbon. In our experiments, we have shown that in comparison to single-laser ablation, the combination of high ion flux and high energy associated with the dual-laser ablation process produces DLC films with a high content of sp^3-hybridized carbon. This result, in conjunction with the ability to grow particulate free films over an extended area, clearly demonstrates the advantages offered by the dual-laser process for the growth of high-quality DLC films.

ACKNOLEDGEMENTS

This research was supported in part by the National Science Foundation (Grant No. DMI-9978738) and the US Department of Energy (Grant No. DE-FG02-96ER12199).

REFERENCES

1. P Mukherjee, J. B. Cuff, and S. Witanachchi, Appl. Surface Sci. **127-129**, 620 (1998).
2. F. S. Pool, Y. H. Shing, J. Appl. Phys. **68**, 3276 (1984).
3. O. Matsumoto, H. Toshima, and Y. Kanzaki, Thin Solid Films **128**, 341 (1985).
4. M. A. Capano, N. T. McDavitt, R. K. Singh, F. Qian, J. Vac. Sci. Technol. A**14**, 431 (1996).
5. A. A. Voevodin and M. S. Donley, Surface and Coating Technology **82**, 199 (1996).
6. Y. Namba and T. Mori, J. Vac. Sci. Technol. A**3**, 319 (1985).
7. C. Weissmantel, K. Bewilogua, D. Dietrich, H. J. Erler, H. J. Hinneberg, S. Klose, W. Nowiek, and G. Reisse, Thin Solid Films **72**, 19 (1980).
8. S. Ravi, P. Silva, S. Xu, B. X. Tay, H. S. Tan, and W. I. Milne, Appl. Phys. Lett. **69**, 491 (1996).
9. C. Weissmantel, Thin Solid Films **58**, 101 (1979).
10. V. I. Merkulov, D. H. Lowndes, G. E. Jellison, A. A. Puretzky, and D. B. Geohegan, Appl. Phys. Lett. **73**, 2591 (1998).
11. A. A. Voevodin, S. J. P. Laube, S. D. Walck, J. S. Solomon, M. S. Donley, and J. S. Zabinski, J. Appl. Phys. **78**, 4123 (1995).
12. S. Witanachchi, Y. Ying, A. M. Miyawa, and P. Mukherjee, Mat. Res. Soc. Symp. Proc. **485**, 185 (1997).
13. S. Witanachchi, K. Ahmed, P. Sakthivel, and P. Mukherjee, Appl. Phys. Lett. **66**, 1469 (1995).
14. S. Witanachchi and P. Mukherjee, J. Vac. Sci. Technol. A**13**, 1171 (1995).
15. J. Wagner, M. Ramsteiner, C. Wild, and P. Koidl, Phys. Rev. B **40**, 1817 (1989).
16. J. Shiao, and R. W. Hoffman, Thin Solid Films **283**, 145 (1996).

Mat. Res. Soc. Symp. Vol. 617 © 2000 Materials Research Society

Optical Properties of Tantalum Oxide Films Deposited on BK7 Substrates by Excimer Laser Ablation

S. Boughaba and M. U. Islam

National Research Council Canada
Integrated Manufacturing Technologies Institute, 800 Collip Circle, London, Ontario
Canada N6G 4X8

ABSTRACT

Thin amorphous films of tantalum oxide were grown on borosilicate crown glass substrates by KrF excimer pulsed laser ablation of a Ta_2O_5 target, in an oxygen environment. The deposition was performed at a temperature of 250 or 400 °C, while the oxygen pressure was set in the range 5 to 30 mTorr. The optical properties of the tantalum oxide coatings, as evaluated by reflectance/transmittance spectrophotometry, were found to be dependent on the oxygen gas pressure. At a pressure of 5 mTorr, absorbing films were obtained, with extinction coefficients above 10^{-2} (at λ=633 nm), along with an optical energy band-gap as low as 0.7 eV. At a pressure of 10 mTorr and above, the coatings had refractive indices up to 2.25 (at λ=633 nm), extinction coefficients below 10^{-4} (for λ>390 nm), and an optical energy band-gap in the range 3.9 to 4.0 eV.

INTRODUCTION

The unique optical properties of tantalum pentoxide (Ta_2O_5) make this material an important candidate for many applications. Indeed, the high refractive index of Ta_2O_5 (n~2.2 at λ= 633 nm), its large band-gap (E_G~4.2 eV), and its low absorption for wavelengths ranging from 300 nm up to 2.0 µm [1-3], are properties that resulted in the emergence of tantalum pentoxide as a key material to be used in photovoltaic, optic, and photonic devices. Ta_2O_5 thin films are used as antireflective coatings for solar cells and charge-coupled devices [3-5]. They are of great interest for optoelectronic applications as propagation and waveguiding channels [6]; in optical devices as birefringent coatings and as components of multilayer interference filters [7, 8]; and for nonlinear optical applications [9].

Numerous investigations of the optical properties of Ta_2O_5 coatings have been reported, in which various chemical and physical deposition methods were employed. Such techniques include electron-beam evaporation [2, 7, 8], reactive sputtering [4, 6], ion-beam sputtering [10], atomic layer deposition [11], reactive low-voltage ion plating [12], sol-gel spin coating [9, 13], and pulsed laser deposition (PLD) [14-16]. The tantalum oxide films obtained by different techniques exhibit variations in their refractive index and extinction coefficient. In our previous work on the synthesis and characterization of tantalum oxide films by PLD on silicon substrates, we demonstrated, using spectral reflectance measurements, that laser ablation allows the deposition of Ta_2O_5 coatings with superior optical properties [15, 16]. In order to complement this investigation, it was necessary to further characterize the coatings by spectral transmittance, using transparent substrates. In the literature, there is no thorough report on simultaneous transmittance and reflectance characterization of tantalum oxide films deposited by laser ablation.

In this article, we present our work on the combined spectral transmittance and reflectance characterization of tantalum oxide films deposited on borosilicate crown glass (BK7) substrates by KrF excimer laser ablation of a Ta_2O_5 target in oxygen (O_2). The refractive index, extinction coefficient, dispersion, and optical energy band-gap are reported.

EXPERIMENT

The deposition of the tantalum oxide films was performed using an advanced PLD setup that is schematically shown in Figure 1. It was previously described in detail [15, 16]. The setup consisted of a KrF excimer laser (λ=248 nm, *Lambda Physik*, LPX-210i) as the ablation energy source, coupled to a PLD system (*Epion Corp.*, PLD-3000) through a variable attenuator (10 to 90% transmission, *Microlas*) and beam steering and shaping optics. The high-vacuum chamber of the system was equipped with a loadlock transfer chamber and was pumped down to, and maintained at, a base pressure below 1×10^{-7} Torr using a turbo-molecular pump. The pulsed laser beam was focused onto a 90 mm diameter rotating Ta_2O_5 target (vacuum hot-pressed, stoichiometric, 99.99%, *Target Materials Inc.*). The laser beam spot size at the target surface was about 2×1 mm^2. The on-target laser beam fluence was adjusted to about 5 J/cm^2, with a repetition rate of 100 Hz. The interaction of the laser with the target produced a plume normal to the target surface and the ejected species were deposited onto a rotating BK7 substrate (25 mm diameter, 0.8 mm thick, polished both sides, 60-40 scratch-dig, flatness < λ/4), located 11.5 cm from the target surface. To obtain uniform films, the laser beam was rastered over a radius of the rotating target using a programmable kinematic mount for the last mirror of the optical train.

Before loading a substrate into the deposition chamber, it was cleaned using methanol. After loading the substrate, the controller of a radiation-based heater was turned on. Quartz lamps on both sides (top and bottom) of the substrate were used as heating elements, allowing non-contact heating. The input to the controller that drove the power to the heating elements was given by a thermocouple, placed above the substrate (no contact). A readout permitted the monitoring of the temperature. When the process temperature was reached and became steady, ultrahigh purity O_2 gas (99.995%, *Air Liquide*) was introduced. The inlet flow of O_2 was set at 3 sccm using a mass-flow controller and deposition pressure was adjusted by throttling the inlet of the turbo-molecular pump. The laser was then turned on and a precleaning of the target was performed for two minutes. Subsequently, the shutter that hid the substrate surface from the ablation plume was opened and the deposition started. After a given processing time, the laser was turned off, the oxygen inlet valve was shut off, and the substrate was allowed to cool down under vacuum.

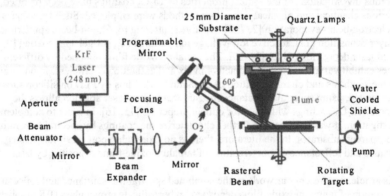

Figure 1 : Schematic of the PLD setup (lower set of quartz lamps not shown).

In the present experiments, tantalum oxide films were deposited at a temperature of 250 and 400 °C. Oxygen gas pressures in the range 5 to 30 mTorr were used. These temperatures and gas pressures were previously found, in the spectral reflectance study of tantalum oxide coatings on silicon substrates, to embrace the processing window that leads to stoichiometric Ta_2O_5 [15, 16].

The structure of the deposits was examined by X-ray diffraction (XRD, *Philips*, X-Pert MRD) in the θ_0-2θ thin film configuration, where θ_0 was fixed at a value of 1°. The thickness, refractive index and extinction coefficient were simultaneously evaluated by spectral reflectance and transmittance, with wavelengths in the range 300 to 850 nm and 420 to 1000 nm, respectively. A normal incidence fiber-optic-based spectrophotometer (*Scientific Computing International*, Filmtek 3000) was used. A generalized Lorentz oscillator model, developed by *SCI*, was used for fitting the reflectance and transmittance curves, and the calculation of the thickness and the optical parameters. This model allows for coupling between oscillators. Two oscillators were used to obtain the best fitting. The thickness of the coatings on control samples was also measured using a stylus profiler (*Veeco*, Dektak 3ST). The values of the thickness obtained by both techniques were consistent. The optical energy band-gap of the deposited films was evaluated using the Tauc-Lorentz model provided by *SCI*.

RESULTS AND DISCUSSION

Tantalum oxide films having a thickness in the range 100 to 130 nm were grown for this study. For all processing conditions, the films were highly adherent to the substrate as revealed by qualitative adhesive tape peel tests. The X-ray diffraction patterns consisted of a diffuse-scattering curve, similar to the ones previously obtained with coatings grown on silicon substrates [15, 16], indicating an amorphouslike structure.

Figure 2 shows the reflectance and transmittance spectra of two different Ta_2O_5 coatings deposited at a temperature of 400 °C and an O_2 gas pressure of 5 mTorr (Figure 2-a) and 10 mTorr (Figure 2-b). The reflectance and transmittance of the bare BK7 substrate are indicated by R_0 and T_0, respectively. At an O_2 pressure of 5 mTorr (Figure 2-a), within the overlapping wavelength analysis region (λ=420-850 nm), the sum of the transmittance, T, and reflectance, R, of the coated substrate is less than one. This means that the tantalum oxide coating deposited at an O_2 pressure of 5 mTorr is absorbing, with an absorbance A, such that R+T+A=1. At an O_2 pressure of 10 mTorr (Figure 2-b), as for all other higher pressures, the sum of the reflectance and transmittance of the coated substrate is equal to 1 within the overlapping wavelength analysis region. Hence, for an O_2 pressure in the range 10 to 30 mTorr, the tantalum oxide films are non-absorbing within this region. A complementary way to determine the absorbing or non-absorbing nature of homogeneous films is to examine the characteristics of the spectral curves at the half-wave points (extrema) of the transmission spectra [8]. At an O_2 pressure of 5 mTorr, the transmission of the coated substrate at the maximum of the spectrum is lower than that of the bare BK7 (T_0), while at a pressure of 10 mTorr these transmissions are equal, revealing the absorbing and non-absorbing nature of the tantalum oxide films, respectively. Similar spectra and behavior, with respect to the O_2 pressure, were observed with the coatings grown at a temperature of 250 °C.

In our previous investigation on the synthesis of tantalum oxide films on silicon substrate, the Ta/O ratios for O_2 gas pressures in the range 1 to 40 mTorr were measured using Rutherford backscattering spectrometry (RBS) and Auger electron spectroscopy [15, 16]. It was found that increasing the O_2 pressure resulted in an increase of the oxygen content of the films, with steady Ta/O ratios in the range 0.38 to 0.44 (considered as stoichiometric Ta_2O_5) for a gas pressure

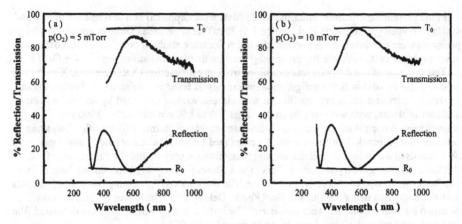

Figure 2 : Reflectance and transmittance spectra of tantalum oxide films deposited at a temperature of 400 °C and an O_2 pressure of (a) 5 mTorr and (b) 10 mTorr. R_0 and T_0 are the reflectance and transmittance of the bare BK7 substrate, respectively.

above 10 mTorr and deposition temperatures up to 400 °C. At lower pressures, oxygen-deficient films were obtained, with Ta/O ratios of up to 0.7. Based on these results, the difference in spectral absorption, as a function of O_2 pressure, of the coatings deposited on BK7 (Figure 2) can be directly related to the oxygen supply during the film growth and oxygen content of the coatings. Stoichiometric Ta_2O_5 coatings are non-absorbing in a wide range of the optical spectrum considered. The absorption of the films grown at the lowest pressure could in addition be related to the generation of defects by the energetic impinging ablated species.

The dependence of the refractive index, n, and extinction coefficient, k, on the wavelength is presented in Figure 3 for tantalum oxide films deposited at a temperature of 250 °C and an O_2 pressure of 5 and 20 mTorr. For both pressures, the coatings were dispersive, with the index of

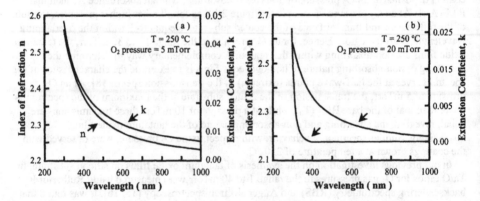

Figure 3 : Dispersion curves of the refractive index and extinction coefficient of tantalum oxide films grown at a temperature of 250 °C and an O_2 pressure of (a) 5 mTorr and (b) 20 mTorr.

refraction decreasing with increasing the wavelength. Such a behavior is representative of that observed with the coatings grown at all temperatures and O_2 pressures, and it is similar to previously reported dispersion data on tantalum oxide films [8, 9, 11, 13]. Moreover, all oxygen-deficient coatings, such as the one considered in Figure 3-a, were found to be absorbing throughout the spectral region investigated. Conversely, as it is the case for the coating deposited at an O_2 pressure of 20 mTorr (Figure 3-b), all stoichiometric Ta_2O_5 coatings were virtually absorption-free ($k<10^{-4}$) for wavelengths in the range 390 to 1000 nm. The sharp increase of the extinction coefficient for shorter wavelengths is associated with fundamental band-gap absorption in the films.

Figure 4 shows the dependence of the refractive index, n, and extinction coefficient, k, on the O_2 pressure of tantalum oxide films, at a wavelength $\lambda=633$ nm, for the two deposition temperatures. At the pressure of 5 mTorr, indices of refraction and extinction coefficients of an average value of 2.26 and 0.015 were obtained, respectively. At pressures of 10 mTorr and above, the indices of refraction were found to be in the range 2.10 to 2.25, with extinction coefficients of less than 10^{-4}. As discussed above, the behavior of the extinction coefficient with the O_2 pressure can be attributed to the oxygen content of the coatings; at a pressure of 5 mTorr, oxygen-deficient films are deposited. Overall, the high values of the index of refraction are indicative of highly packed, dense films, and are among the highest refractive indices reported for Ta_2O_5 [3, 10, 13]. At the highest pressure of 30 mTorr, the value of the index of refraction was found to be lower than that obtained at other pressures. This could be attributed to the reduction of the energy of the ablated tantalum-bearing species, which undergo several collisions with the background O_2 molecules before reaching the growing film surface. Such low-energy species result in a lower density film with smaller value of the refractive index.

The optical energy band-gap, E_G, of the tantalum oxide films was investigated as a function of oxygen pressure. Figure 5 shows this dependence for the two deposition temperatures, 250 and 400 °C. The energy band-gap of the coatings deposited at an O_2 pressure of 5 mTorr had an average value of 0.8 eV, increasing to a steady value in the range 3.9 to 4.0 eV for pressures of 10 mTorr and higher. According to the discussion above, the increase of the energy band-gap with the O_2 pressure could be associated with the higher oxidation level of the films. The values

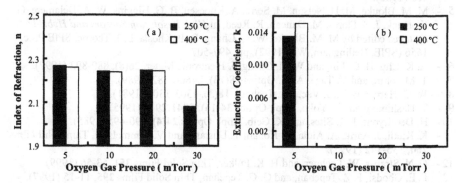

Figure 4 : Dependence of the index of refraction (a) and extinction coefficient (b) of tantalum oxide films on the O_2 pressure. Values of n and k are given for $\lambda=633$ nm.

of the energy band-gap of stoichiometric films (O_2 pressure ≥ 10 mTorr) are similar to the highest values (4.0-4.5 eV) reported for Ta_2O_5 films [10, 13].

CONCLUSIONS

Amorphous tantalum pentoxide films, with a high refractive index (2.1<n<2.25 at λ=633 nm), a low extinction coefficient (k<10^{-4} for λ>390 nm), and a large optical energy band-gap ($3.9<E_G<4.0$ eV) were successfully deposited on BK7 substrates within a wide processing window. Such characteristics demonstrate that high-quality Ta_2O_5 films, eligible for optical-based applications, can be produced by PLD at relatively low temperatures.

ACKNOWLEDGEMENTS

The authors are indebted to Mr. Michael Meinert for his technical help.

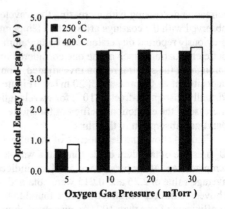

Figure 5 : Dependence on the O_2 pressure of the optical energy band-gap of tantalum oxide coatings deposited at a temperature of 250 and 400 °C.

REFERENCES

1 - P. J. Harrop and D. S. Campbell, Thin Solid Films 2, 273-292 (1968).

2 - J. D. Traylor-Kruschwitz and W. T. Pawlewicz, Appl. Optics 36(10), 2157 (1997).

3 - C. Chaneliere, J. L. Autran, R. A. B. Devine, and B. Balland, Materials Science and Engineering R22, 269-322 (1998).

4 - F. Rubio, J. Denis, J. M. Albella, and J. M. Martinez-Duart, Thin Solid Films 90, 405-408 (1982).

5 - M. M. Blouke, M. D. Nelson, M. Serra, A. Knoesen, B. G. Higgins, W. A. Delamere, G. Womack, J. S. Flores, M. Duncan, R. Reed, in High Resolution Sensors and Hybrid Systems, edited by M. M. Blouke, W. C. Chang, R. P. Khosla, L. J. Thorpe, SPIE Proc. 1656 (SPIE, Bellingham, WA, 1992), pp. 497-507.

6 - A. K. Chu, H. C. Lin, and W. H. Cheng, J. Electronic Mater. 26(8), 889-892 (1997).

7 - T. Motohiro and Y. Taga, Appl. Optics 28(13), 2466-2482 (1989).

8 - W. C. Hermann, Jr., J. Vac. Sci. Technol. 18(3), 1303-1305 (1981).

9 - T. Hashimoto and T. Yoko, Appl. Optics 34(16), 2941-2948 (1995).

10- H. Demiryont, J. R. Sites, and K. Geib, Appl. Optics 24(4), 490-495 (1985).

11- K. Kukli, J. Aarik, A. Aidla, O. Kohan, T. Uustare, and V. Sammelselg, Thin Solid Films 260, 135-142 (1995).

12- G. N. Strauss, W. Lechner, and H. K. Pulker, Thin Solid Films 351, 53-56 (1999).

13- F. E. Ghodsi, F. Z. Tepehan, and G. G. Tepehan, Thin Solid Films 295, 11-15 (1997).

14- J.-Y. Zhang, Q. Fang, and I. W. Boyd, Appl. Surf. Sci. 138-139, 320-324 (1999).

15- S. Boughaba, M. U. Islam, G. I. Sproule, and M. J. Graham, Surface & Coatings Technology 120-121, 757-764 (1999).

16- S. Boughaba, G. I. Sproule , J. P. McCaffrey, M. Islam, and M. J. Graham, Thin Solid Films, 358(1-2), 104-113 (2000).

Mat. Res. Soc. Symp. Vol. 617 © 2000 Materials Research Society

Investigation on Laser-Induced Effects in Nanostructure Fabrication with Laser-Irradiated Scanning Tunneling Microscope Tips in Air Ambient

Z. H. Mai, Y. F. Lu, W. D. Song, and W. K. Chim
Laser Microprocessing Laboratory, Department of Electrical Engineering and Data Storage Institute, National University of Singapore, 10 Kent Ridge Crescent, Singapore 119260

ABSTRACT

In this paper, we report our investigation on the kinetics of nanostructure fabrication on gold films and on H-passivated Ge surfaces. The relationship between the current and the tip-sample distance of the STM junction was measured for both gold films and H-passivated Ge surfaces. The tip-sample distance for gold films under a electrochemically etched W tip is approximately 2 nm, while that for H-passivated Ge sufaces is more than 27 nm. The thermal expansion length of the tip under laser irradiation was calculated. From the comparison of the thermal expansion length and the tip-sample distance, we can reach the conclusion that for gold films, thermal mechanical indention is the primary reason of nanostructure formation, while for H-passivated Ge surfaces, optical enhancement is the only reason.

1. INTRODUCTION

Various techniques have been developed for nanostructure fabrication since the invention of the STM. Applications of the laser-assisted STM arouse wide interest after Cutler, et al., [1] proposed the use of a laser-STM combination for the measurement of a tunneling time and suggested possible laser-STM studies. One of the promising applications of the laser-assisted STM is nanostructure fabrication. Recently, several researches on the nanostructure fabrication using pulsed lasers in combination with an STM have been reported. [2-10] In this paper, we investigate mechanism of nanostructure fabrication on gold films and H-passivated Ge and Si surfaces using lasers in combination with an STM. Current-distance curves for tip-sample junctions were measured. An analytical model was established to distinguish the dominant mechanism.

2. EXPERIMENTAL

Gold films with a thickness of 100 nm were deposited with physical vapor deposition. p-type Ge (100) wafers with a resistivity of 8-12 Ωcm were H-passivated in a 50% HF solution. n-type Si (100) wafer with a resistivity of 10-30 Ωcm were H-passivated in a 5% HF solution. Electrochemically etched STM tips were homemade from a 0.5 mm W wire

Nanostructures, such as dots and lines, were created on gold films and H-passivated Ge and Si surfaces. The dependence of the height or depth of the nanostructures was measured. Current-distance curves of the STM junctions were measured and compared with calculated thermal expansion of the tip.

In the air ambient, thin water layer forms on any solid surface.[11] Water layers will influence the tip-sample distance for STM imaging. The curves of current versus tip-sample distance for an EC etched W tip on a gold film and on an H-passivated Ge surface were measured in air at a relative humidity of 88%. The tip, firstly, was automatically approached the sample surface at a preset current value, which correspond to the maximum current value in our measurement. During auto-approach, the feedback was on. The scanner stopped when the current reached the preset value. Then the feedback was disabled and the scanner was retracted a little to reduce the current to zero. The current-distance curve for extending process of the scanner was measured. After the preset current value was reached, the scanner was stopped and then retracted, and the current-distance curve was measured for retracting process. We retracted or extended the scanner by a step of 1 nm, which is the minimum manually adjustable step in our system, and the average current was recorded for different scanner position. The average current was measured by averaging the current signal recorded in 1 minute using the built-in multi-meter. After the measurement of the current-distance curves, STM imaging was taken to examine the sample surface. A preset value of 10 nA and 3 nA were used for the measurement on gold films and H-passivated Ge surfaces, respectively. The preset current value must be low enough to make sure that no surface damage or modification occurs.

3. RESULTS

Figure 1 (a) is a STM image of a 2 × 2 oxide dot array created by a single laser pulse for each dot under an EC etched tungsten tip. The intensity is approximately 3 MW/cm^2. STM imaging was taken using the same tip after nanoprocessing. The slight differences in the dot sizes and apparent depths are due to the pulse to pulse fluctuation of laser output energy. The spectroscopy (STS) on the processing regions were studied before and after laser processing. The STS studies showed a decreased conductivity in the processing regions after laser processing. These dots are with a size between 20 nm to 30 nm, and with an apparent depth of approximately 8 nm. The shape of dots depends on the geometry of the tip. Although the dots appear as depressed regions in the STM image, they appear as mounds in an AFM image of the same region, as shown in Fig. 1 (b). The AFM image shows that the height of the dots was approximately 0.9 nm.

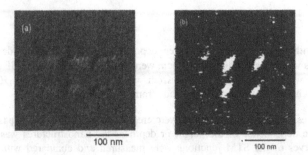

Figure 1. A 2 × 2 dot array of oxide created by a tungsten STM tip under laser irradiation: (a) STM image and (b) AFM image.

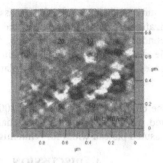

Figure 2. Nano pits created at different laser intensity.

Figure 2 is an STM image of 5 dots created at different laser intensity. In Fig. 4-1, we can see that these dots have a feature of craters.

Figure 3 (a) is the current-distance curve for gold films. The current-distance curves were measured for both a pre-cut W tip and an EC etched W tip. No obvious difference in current-distance curves for a pre-cut W tip or an EC etched W tip was observed. A transition length from 0 to 10 nA was longer than 2 nm and shorter than 3 nm. Obviously, the gold surface is too hydrophobic to adsorb sufficient water, resulting in a very thin water layer and a short water bridge, which can be maintained between the W tip apex and the gold sample. However, 2 or 3 nm is also larger than the tunneling regime in vacuum. Therefore, the current is partially contributed by the conductivity of the water bridge. No hysteresis of the current-distance curve was observed for gold film. This is probably because of the large retracting or extending step.

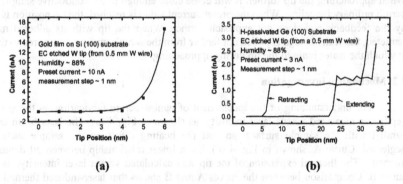

(a)   (b)

Figure 3. Current-distance curves for STM junctions with an EC-etched W tip on (a) gold films and (b) H-passivated Ge (100) surfaces.

Figure 3 (b) is the current-distance curve for H-passivated Ge (100) surfaces. The curve shows different characterization. A plateau appears in the curve. The mechanism of the STM junction is based on the conductivity of the water bridge. After the bridge is built up, with the decrease of the tip-sample distance, the current maintain constant until the two water layer contact, the current increases rapidly. The measured plateau width for retracting process is approximately 27 nm.

The current-distance curve for H-passivated Si (100) surfaces was also measured using the above mentioned method. The curve shows the same characterization as that for H-passivated Ge (100) surfaces. The measured plateau width for retracting process is approximately 19 nm.

4. DISCUSSION

4.1. STM tip-sample junctions in air ambient

The tip-sample junction in an STM can be modeled to a metal-vacuum-metal (or semiconductor) structure in vacuum, while electron tunneling occurs between the two conductive electrodes. The tunneling current drops by nearly an order of magnitude for every 1-angstrom. Thus such tunneling can be observed in practice only for very small separations, usually approximately 1 nm or so. However, in air ambient, due to the humidity, the tip-sample junction is totally different. On both the surfaces of the tip and the sample, there is a molecular thin water film adsobed (for example, approximately 0.75 nm at 90 % relative humidity on mica with an atomic smooth surface). On surfaces of near smooth substrates, such as silicon, germanium, and films deposited on near smooth substrates, porosity, roughness, and chemical properties will actually affect the thickness of the adsorbed water film.

When the tip is approached towards the sample surface, it first comes into contact with the conductive water layer. This results in a current, which is dependent on the applied voltage. When approaching the tip further, it will come close enough to the conductive sample surface to permit ordinary tunneling. When the preset current value is reached, the tip position is controlled by the feedback and does not approach further. When the tip with its adhered meniscus is retracted, the water bridge is pulled out farther from the sample surface, compared to the distance at which the water jump to the tip during approaching.

4.2. Mechanism investigation

For understanding of the mechanism of nanostructure fabrication on the gold film, pit depth was measured versus laser intensity, as shown in Fig. 4 (a). The laser beam is almost in parallel with the sample surface, so that the heating effect on the sample surface can be neglected. Curve A, shown in Fig. 4 (a), has a linear relationship between pit depth and laser intensity. The thermal expansion of the tip was calculated versus laser intensity, as shown by curve B. Comparison between the curves A and B shows that laser-induced thermal expansion cannot be neglected for nanosecond pulsed laser and the mechanism is based on thermal mechanical indention with the laser heated STM tips.

Figure 4 (b) shows the dependence of the oxide apparent depth on laser intensity (curve A) and the calculated thermal expansion of the tip length (curve B). From the calculation, we can see that during nanooxdation underneath the tip apex, the thermal expansion is much less than the plateau width shown in Fig. 3 (b). This precludes the penetration of tip into the sample. During the thermal process of tip under the laser irradiation, the current-distance relationship is in the plateau region. Therefore, there is no current increase in the tip-sample junction. The mechanism of oxidation on H-passivated Ge surfaces is due to desorption of hydrogen atoms.

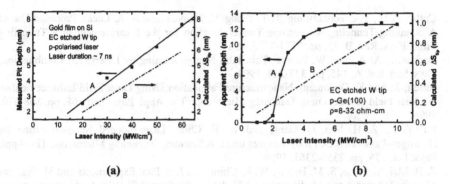

(a)   **(b)**

Figure 4. (a) Pit depth on gold films and (b) oxide apparent depth on H-passivated Ge surfaces versus laser intensity (curve A) and thermal expansion length of the tip versus laser intensity (curve B).

The binding energy of Ge-H bonds on H-passivated Ge surfaces is in the range of 4.6-4.9 eV. [12] The photon energy of a 532 nm laser cannot directly break the Ge-H bonds. Under laser irradiation, the hydrogen atoms were thermally desorbed. After hydrogen desorb from the surface, O- and OH- groups, replacing the adsorbed hydrogen atoms and resulting in oxidation. The original laser irradiation and the tip-enhanced optical field contribute the thermal process on the sample surface underneath the tip.

5. CONCLUSION

In summary, the mechanism of nanostructure fabrication on gold films and H-passivated Ge and Si surfaces was investigated in air ambient. An analytical model was used to discuss the mechanism of the nanostructure formation. On gold films, the tip-sample distance is approximately 2 nm, transient expansion of the tip under laser irradiation enable the tip penetrating the sample. Thermal mechanical indentation created the pits on the gold films. On H-passivated Ge and Si surfaces, the STM junction has a large tip-sample distance, resulting in a wide plateau in the current-distance curve. During modification, the thermal expansion is smaller than the tip-sample distance and there is no current increase. The original laser irradiation and the tip-enhanced optical field thermally desorbed the hydrogen atoms from the sample surface, resulting in oxidation. When the laser beam is in parallel with the sample surface, the mechanism of nanoxidation is based on the optical enhancement underneath the tip.

ACKNOWLEDGEMENT

The authors thank Ms. H. L. Koh and Mr. Y. W. Goh for their kind help in installation of the SPM system and other facilities in our experiments. The work was supported by National University of Singapore under grant No. RP3972692.

REFERENCES

1. Cutler, P. H., T. e. Feuchtwang, T. T. Tsong, H. Nguyen, and A. A. Lucas. Proposed Use of A Scanning-Tunneling-Microscope Tunnel Junction for the Measurement of a Tunneling Time, Phys. Rev. B, 35, pp. 7774-7775. 1987.
2. Gorbunov, A. A., and W. Pompe. Thin Film Nanoprocessing by Laser/STM Combination, Phys. Stat. Sol. A, 145, pp. 333-338. 1994.
3. Jersch, J., and K. Dickmann. Nanosrtucture Fabrication Using Laser Field Enhancement in the Near Field of a Scanning Tunneling Microscope Tip, Appl. Phys. Lett., 68, pp. 868-670. 1996.
4. Lu, Y. F., Z. H. Mai, Q. Gang, and W. K. Chim. Laser-Induced Nano-Oxidation on Hydrogen-Passivated Ge (100) Surfaces under A Scanning Tunneling Microscope Tip, Appl. Phys. Lett., 75, pp. 2359-2361. 1999.
5. Z. H. Mai, Y. F. Lu, S. M. Huang, W. K. Chim, and J. S. Pan. Experiments and Mechanism of Laser-Induced Nano Modification on Hydrogen-Passivated Si (100) Surfaces underneath the Tip of a Scanning Tunneling Microscope. Submitted to J. Appl. Phys.
6. Ukrainstev, V. A., and J. T. Yates, Jr. Nanosecond Laser Induced Single Atom Deposition with Nanometer Spatial Resolution Using A STM, J. Appl. Phys. 80, pp. 2561-2571. 1996.
7. Boneberg, J., H. J. Munzer, M. Tresp, M. Ochmann, P. Leiderer. The Mechanism of Nanostructureing upon Nanosecond Laser Irradiation of a STM Tip, Appl. Phys. A, 67, pp. 381-384. 1998.
8. Lyubinetsky, I., Z. Dohnalek, U. A. Ukraintsev, and J. T. Yates, Jr. Transient Tunneling Current in Laser-Assisted Scanning Tunneling Microscopy, J. Appl. Phys., 82, pp. 4115-4117. 1997.
9. Boneberg, J., M. Tresp, M. Ochmann, H. J. Munzer, and P. Leiderer. Time-Resoled Measurements of the Response of a STM Tip Upon Illumination with a Nanosecond Laser Pulse, Appl. Phys. A, 66, pp. 615-619. 1998.
10. Jersch, J., F. Demming, I. Fedotov, and K. Dickmann. Time-Resolved Current Response of a Nanosecond Laser Pulse Illuminated STM tip, Appl. Phys. A, 68, pp. 637-641. 1999.
11. Heim, M., R. Eschrich, A. Hillebrabd, H. F. Knapp, R. Guckengerger. Scanning Tunneling Microscopy based on the Conductivity of Surface Adsorbed Water. Charge Transfer between Tip and Sample via Electrochemistry in a Water Meniscus or via Tunneling, J. Vac. Sci. Technol. B, 14, pp. 1498-1502. 1996.
12. Larsson, C. U., and A. S. Flodstrom. H_2O Adsorption on Ge (100): An Angle-Resolved Photoelectron Spectroscopy Study, Phys. Rev. B, 43, pp. 9281-9283. 1991.

Mat. Res. Soc. Symp. Vol. 617 © 2000 Materials Research Society

MAGNETIC FIELD GENERATION AT EARLY-STAGE KrF EXCIMER LASER ABLATION OF SOLID SUBSTRATES

M. H. HONG, Y. F. LU, A. FOONG, T.C. CHONG
Laser Microprocessing Laboratory, Department of Electrical Engineering and Data Storage Institute, National University of Singapore, 10 Kent Ridge Crescent, Singapore 119260

ABSTRACT

Magnetic field generation at early-stage of KrF excimer laser ablation of solid substrates (for delay time less than 200 ns) is investigated. Based on classical electrodynamics, fast and dynamic emission of electrons and positive ions at the beginning of laser ablation induces an electromagnetic field nearby. A tiny iron probe wrapped with 50-turn coil is applied to detect this dynamic magnetic field. It is found that signal waveform is closely related to probe distance from the laser spot. For probe distance less than a threshold, the signal has a double-peak profile with a negative peak appearing first and peak durations in tens of nanoseconds. As probe distance increases, the amplitude of positive peak reduces much faster than the negative one. It disappears for probe distance up to the threshold and signal waveform becomes a negative peak with a wider duration. Mechanism on magnetic field generation at early-stage of laser ablation is analyzed to obtain more information on charged particle dynamics. Dependence of the magnetic signal on laser fluence, substrate bias and pulse number is also studied during laser ablation of solid substrates and removal of metallic oxide layer on the surface.

INTRODUCTION

High-power short-pulse laser ablation has been extensively applied in thin film deposition, substrate patterning and surface cleaning [1-3]. It is a fast dynamic process with substrate materials removal, plasma generation and interaction with incident laser light in a short time [4]. Since there are not enough optical and thermal parameters to reflect the process properly, it is very difficult for theoretical modeling to give a clear picture of its dynamics. Recently, many research efforts are carried out to analyze the acoustic, electric, magnetic and optical signals in order to have a better understanding of laser ablation [5-8]. For examples, piezoelectric sensor was applied for surface vibration, time-of-flight (TOF) spectroscopy for charged particle behaviors and optical multichannel analyzer (OMA) for plasma evolution. However, these techniques have disadvantages of complicated setup, direct coupling of sensor and substrate and difficulty to get information on early-stage for delay time less than 200 ns [9]. It is a challenge to develop non-contact and fast detection to study laser ablation. In our previous study, a tiny metal probe was used as an antenna to detect plasma-induced electric field [10]. Based on classical electrodynamics, electrons and positive ions emitted at very beginning of laser ablation induce an electromagnetic field nearby [10]. From signal waveform, information on charged particle emission can be referred. Further to the study, magnetic signal generation at the early-stage is studied in this work. Signal variation with laser fluence, probe distance and substrate bias is analyzed. Signal diagnostic is also applied to monitor pulsed laser removal of metallic oxide layer in real time.

EXPERIMENT

A KrF excimer laser (Lambda Physik, LPX 100) was used as a light source. Laser beam has a wavelength of 248 nm and a pulse duration (FWHM) around 23 ns. After passing through a beam splitter, 5% of laser energy was irradiated into an ultrafast phototube (Hamamatsu R1328U-53, rise time 60 ps and fall time 55 ps) to capture laser pulse profile. The other part of laser energy was focussed onto a substrate surface by a quartz lens with a focal length of 150 mm. Laser fluence was adjusted by varying output energy. A polycrystalline copper disk was used as substrate with its surface perpendicular to incident laser beam. It was placed on an X-Y stage to change laser irradiation position on the surface after signal detection and offer a fresh area for next laser pulse. The substrate was connected to the ground during signal detection. A tiny iron probe (tip diameter: 0.1 mm) wrapped with 50-turn coil was applied to detect magnetic signal. Its position was tuned by a micrometer. Dynamic magnetic field passing through the iron probe induces a current inside the coil. It was then sent to a fast digital oscilloscope (Lecroy LC 534A, bandwidth 1 GHz with sampling rate 2 GS/s). Synchronous output of laser controller was sent to the oscilloscope as a trigger signal for magnetic signal and laser pulse detection. Digitized signal was sent to a PC through an IEEE-488 interface for data storage and further processing. The signal detection was also applied in real-time monitoring of laser removal of tungsten oxide layer. A power supply with constant voltage output from -110 to +110 V was used to apply a substrate bias and study the signal variation. A tiny metal probe (diameter: 0.2 mm) was used to detect the electric signal at the same time for comparison with the magnetic one.

RESULTS AND DISCUSSION

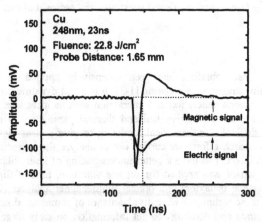

Fig. 1 Electric and magnetic signals recorded at the same time during KrF excimer laser ablation of copper substrate at a laser fluence of 22.8 J/cm^2 and a probe distance of 1.65 mm.

Figure 1 shows electric and magnetic signals recorded at the same time during KrF excimer laser ablation of copper substrate at a laser fluence of 22.8 J/cm^2 and a probe distance of 1.65 mm. The electric signal is moved down 75 mV for a good view of comparison with the magnetic one. It is clear that the magnetic signal is in a dual-peak structure. A negative peak with the sharper profile and higher peak amplitude appears at a delay time of 15 ns with respect to the starting of laser irradiation. It reaches the maximum at a delay time of 20 ns and then rises back to zero 10 ns later. As time increases, the magnetic signal increases up to a positive maximum at a delay time of 37 ns and then reduces to zero about 62 ns later. It can also be observed from Fig. 1 that there is only one negative peak for the electric signal, which appears at the same time as the magnetic signal. Meanwhile, its peak arrival time is at the same time point as the magnetic signal changes from negative to positive.

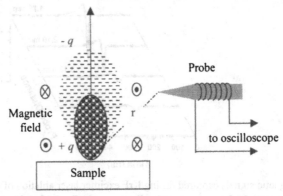

Fig. 2 Magnetic field generation during pulsed laser ablation of solid substrate. Dynamic magnetic field passes through the probe and induces a current inside the coil.

Figure 2 illustrates magnetic field generation during pulsed-laser ablation of solid substrate. At very beginning of laser ablation, electrons with varying speed and concentration are emitted out first in a direction normal to the substrate surface. The magnetic field moves in a circle around the electrons and goes into the paper at left-hand side and out of the paper at right-hand side. It passes through an iron probe nearby and induces a current inside the coil. Magnetic field dynamics can be obtained from the current signal. Starting time of its waveform corresponds to onset of electron emission. As time increases, more and more electrons are emitted and generated a stronger magnetic field. Since speed of positive ions is much lower, their induced magnetic field is smaller. It moves in the opposite direction and appears with a time delay. Therefore, electron emission dominates magnetic field generation at the very beginning. As time increases further, electrons are emitted completely with the maximum signal amplitude. It can be inferred that its peak arrival time corresponds to total emission of electrons. It is also starting time for positive ion emission since signal amplitude starts to decrease. As the process proceeds, electron dynamics reduces while positive ion emission increases. It causes the signal to increase gradually to zero, which corresponds to total emission of positive ions. It is due to neutralization of the magnetic signals induced by all charged particles. From Fig. 1, this instance is at a same time point as the electric signal reaches its maximum amplitude. It also confirms that the electric signal is resulted from an electric dipole constructed by total emitted electrons and positive ions at the early-stage [10]. As the time increases, electrons are moving farther away and positive ion emission dominates the magnetic signal generation. The signal amplitude increases up to a positive maximum and then reduces to zero with plasma expansion and recombination.

Figure 3 demonstrates magnetic signal variation with probe distance during the laser ablation of copper substrate at a laser fluence of 28.9 J/cm^2 and probe distances of 1.76, 2.10, 3.13, 3.59 and 4.06 mm. It can be observed that signal amplitude reduces with probe distance. For the probe distance of 3.13 mm, the positive peak disappears and the negative peak width increases suddenly. It maybe due to a special magnetic field distribution with near-field and far-field at early-stage of laser ablation. In the near field, with probe distance around plasma dimension at the early-stage, both electrons and positive ions make contributions to magnetic field generation.

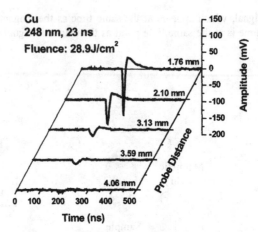

Fig. 3 Magnetic signals captured during KrF excimer laser ablation of copper substrate at a laser fluence of 28.9 J/cm^2 and probe distances of 1.76, 2.10, 3.13, 3.59 and 4.06 mm.

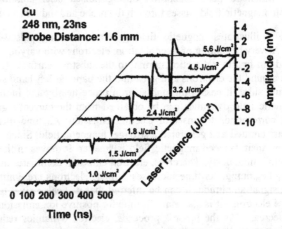

Fig. 4 Magnetic signals detected during KrF excimer laser ablation of copper substrate at a probe distance of 1.6 mm and laser fluences of 1.0, 1.5, 1.8, 2.4, 3.2, 4.5 and 5.6 J/cm^2.

While in the far field with a high probe distance, electron emission dominates the magnetic field generation. As the probe distance increases further, the negative peak reduces to zero gradually. Figure 4 shows the magnetic signals detected during the laser ablation of copper substrate at a probe distance of 1.6 mm and laser fluences of 1.0, 1.5, 1.8, 2.4, 3.2, 4.5 and 5.6 J/cm^2. It can be found that both positive and negative peaks reduce with laser fluence. This is because there is a stronger laser ablation with more electrons and positive ions emitted at a higher laser fluence. For laser fluence of 1.5 J/cm^2, the positive peak disappears and negative peak width increases suddenly. It could be explained by the space distribution and time evolution of the magnetic field at early-stage of laser ablation for a low laser fluence. While for laser fluence of 1.0 J/cm^2, the negative peak also disappears. This is because there are no laser ablation and plasma generation. From the experimental results, threshold fluence for KrF excimer laser ablation of copper substrate can be estimated to be 1.2 J/cm^2.

Magnetic signal detection is also applied to laser ablation induced removal of oxide layer from tungsten substrate. Magnetic signals recorded at a probe distance of 1.6 mm are shown in Fig. 5. Laser fluence applied was 1.3 J/cm². It can be observed that the signal amplitude increases at the first two laser pulses. It is probably because there is less charged particle generation during the laser ablation of the outmost oxide layer. As the pulse number increases, the signal amplitude reduces gradually to zero for pulse number up to 10. Surface morphology comparison before and after the laser processing shows that the oxide layer is completely removed without any damage on tungsten substrate. It means that the magnetic signal diagnostics can be applied in the real-time monitoring of pulsed laser induced removal of metallic oxide layer.

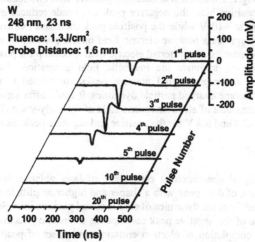

Fig. 5 Magnetic signal detection in the real-time monitoring of pulsed laser removal of tungsten oxide layer for a probe distance of 1.6 mm and a laser fluence of 1.3 J/cm².

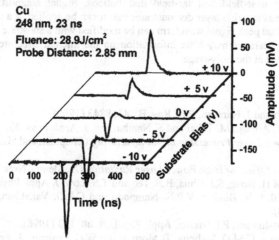

Fig. 6 Magnetic signals captured during KrF excimer laser ablation of copper substrate at a laser fluence of 28.9 J/cm², a probe distance of 2.85 mm and substrate biases of -10, -5, 0, +5 and +10 V.

Experimental results discussed previously were obtained with the substrate connected to the ground and the magnetic signal is resulted from fast dynamics of electrons and ions. Since an external electric field can change behaviors of electrons and positive ions, the magnetic signal can be modified by a substrate bias. Figure 6 presents magnetic signals detected during the laser ablation of copper substrate at a laser fluence of 28.9 J/cm^2, a probe distance of 2.85 mm and substrate biases of -10, -5, 0, +5 and +10 V. For a negative bias, the positive peak reduces gradually and finally disappears while the negative peak amplitude increases greatly with the bias. This is because the negative bias applies a negative electric field on the plasma. It expels the electrons and increases their speed so that the magnetic field induced by the electrons is stronger. However, it attracts the positive ions and reduces their speed gradually to zero. For a positive bias, the negative peak amplitude reduces gradually to zero for the substrate bias up to +10 V while the positive peak increases tremendously. This is because the positive bias applies a positive electric field on the plasma, which attracts the electrons and reduces its dynamics. The external electric field speeds up the positive ions, which causes the positive ions to dominate the magnetic field generation. Threshold biases for the disappearance of the negative and positive peaks are related to minimum electric field intensities to overcome charged particle dynamics. It could offer some important information on the internal structure of electromagnetic field at the early-stage. The threshold biases can be estimated as -2.4 and 6.8 V for the positive and negative peak disappearance, respectively.

CONCLUSIONS

Magnetic signal generation at early-stage of laser ablation is investigated. Magnetic signal is in a form of dual peak with a shaper and higher amplitude negative peak appearing first. It is resulted from the dynamics of electrons and positive ions. Signal analyses show that the starting time of the negative peak corresponds to onset of electron emission and its peak arrival time to completion of electron emission and onset of positive ion emission. Time instance at the signal changing from negative to positive corresponds to total emission of positive ions. Signal variation with probe distance and laser fluence implies that there probably exist near-field and far-field distributions. Signal diagnostic during the laser removal of metallic oxide layer demonstrates that it can be applied as a real-time monitoring scheme. The dual peak signal waveform can be modified with a substrate bias. Threshold bias for peak disappearance may offer information on the internal structure of electromagnetic field constructed at the early-stage.

REFERENCES

1. R.K. Singh and J. Narayan, Phy. Rev. B. **41**, 8843 (1990).
2. Y.F. Lu, Y. Aoyagi, M. Takai and S. Namba, Jpn. J. Appl. Phys. **33**, 7138 (1994).
3. M.C. Gower, *Laser Processing in Manufacturing* (Chapman and Hall, London, 1993), p. 189.
4. J.F. Ready, *Effects of High Power Laser Radiation* (Academic, New York, 1971), p. 127.
5. Y.F. Lu, M.H. Hong, S.J. Chua, B.S. Teo and T.S. Low, J. Appl. Phys. **79**, 2186 (1996).
6. S.S. Harilal, C.V. Bindhu, V.P.N. Nampoori and C.P.G. Vallabhan, Appl. Phys. B. **66**, 633 (1998).
7. A. V. Kabashin and P.I. Nikitin, Appl. Phy. Lett. **68**, 173 (1996).
8. J.M. Hendron, C.M.O. Mahony, T. Morrow and W.G. Graham, J. Appl. Phys. **81**, 2131 (1997).
9. Y.F. Lu and M.H. Hong, J. Appl. Phys. **86**, 2812 (1999).
10. H.C. Ohanian, *Classical Electrodynamics* (Allyn & Bacon, London, 1988), p. 56.

Mat. Res. Soc. Symp. Vol. 617 © 2000 Materials Research Society

Inflence of Substrate Temperature on Barium Ferrite Films Prepared by Laser Deposition

W.D. Song, Y.F. Lu, W.J. Wang, T.C. Chong, Laser Microprocessing Laboratory, Department of Electrical Engineering and Data Storage Institute, 10 Kent Ridge Crescent, Singapore 119260

ABSTRACT

Influence of substrate temperature on properties of barium ferrite films prepared by laser deposition is studied in this paper. The magnetic properties, grain shape and crystalline orientation of the films are discussed for the films prepared by laser deposition with in-situ heating, post annealing and varying substrate temperature. The results show that magnetic properties, grain shape and crystalline orientation of the film deposited with varying substrate temperature are close to the film deposited with post-annealing and different to the film deposited with in-situ heating.

INTRODUCTION

In the last few years, the area density of rigid disk has been increasing at an annual growth rate of 60%. If this growth rate continues, 40 Gb/in^2 drives will be on the market around year 2003 [1]. In order to realize such area density, it is necessary to develop media material with coercivity of 5 kOe and above [1]. Barium ferrite has been identified as one of the candidates for magnetic recording since it possesses both high coercivity and good mechanical and chemical stability [2,3].

Presently, barium ferrite films have been prepared by a few groups using sputtering either applying in situ substrate heating during deposition or through a post annealing process [3-14]. For in-situ heating, substrates are heated up to a certain temperature before deposition starting, then substrate temperature is kept constant during film growth. After that, films are cooled down to room temperature. In contrast, for post annealing, films are grown at room temperature or a low temperature. The films are then processed with a post annealing. Both methods have been widely used in sputtering and laser deposition. For these two methods, the substrate temperature is always constant during film growth. Since substrate temperature is one of the key parameters to the film quality, it is of interest to investigate the influence of varying substrate temperature on film properties. In this paper, we report laser deposition of barium ferrite films with in-situ heating, post-annealing and varying substrate temperature during film growth.

EXPERIMENTAL

The experimental setup of laser deposition system was discussed elsewhere [14]. A KrF excimer laser beam was focused onto a rotating target with a lens to produce a laser fluence of about 3 J/cm^2. A barium ferrite (BaFe$_{12}$O$_{19}$) target was mounted at 45° with respect to the laser beam. Facing the target at a distance of 4 cm from it, (001) single-crystal Al$_2$O$_3$ substrates were mounted on a stainless steel holder by silver paste. A background pressure of 10^{-5} Torr was achieved with a turbomolecular pump. During the deposition, about 250 mTorr pressure of flowing oxygen was maintained. The first group of films was prepared by laser deposition with in-situ heating at 900 °C during film growth. The second group of films was prepared by laser deposition with a post annealing, that is, films deposited at room temperature and followed by annealing at 900 °C in the vacuum chamber at the same pressure of flowing oxygen. The third group of samples was prepared by laser deposition with varying substrate temperature during film growth.

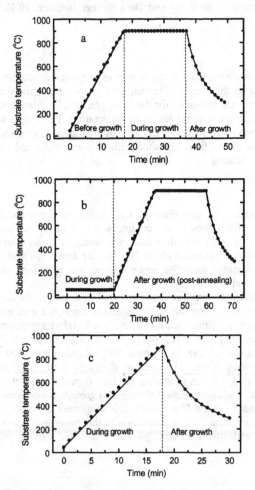

Fig. 1 The profiles of substrate temperature variation
with time before, during and after deposition

RESULTS AND DISCUSSION

Figure 1 shows the profiles of substrate temperature variation with time before, during and after deposition. From Fig. 1(a), it was found that substrate temperature increases with time at a rate of 50 °C/min before film growth, keeps at 900 °C during film growth and cools down to room temperature after film growth. From Fig. 1(b), it was observed that substrate temperature keeps at room temperature during film growth. After film growth, it increases with time at a rate of 50 °C/min, keeps at 900 °C for post-annealing and cools down to room temperature. From Fig. 1(c), it was seen that substrate temperature increases with time at a rate of 50 °C/min during film growth and cools down to room temperature after film growth.

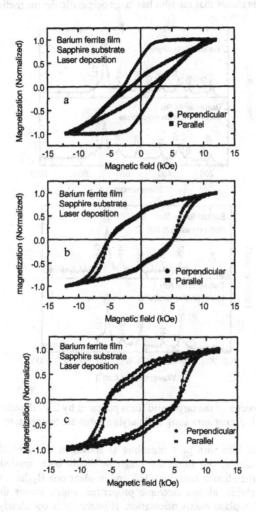

Fig. 2 The magnetization hysteresis loops of barium ferrite films prepared by laser deposition
With (a) in-situ heating, (b) post annealing and (c) varying substrate temperature

The magnetization hysteresis loops of barium ferrite films prepared by laser deposition with in-situ heating, post annealing and varying substrate temperature during film growth are shown in Fig. 2. It is observed that the film with in-situ heating as shown in Fig. 2(a) is easily saturated as magnetic fields are applied perpendicular to the film plane. However, if increasing fields are applied parallel to the film plane, the magnetization increases linearly until the film is saturated at a very large field. This indicates that the easy axis of magnetization is normal to film plane. For barium ferrite, the easy axis is parallel to the c-axis. Therefore, c-axis is verified to be normal to the film plane, which is consistent with the XRD analysis [14]. The perpendicular and in-plane coercivities $H_{c\perp}$ and $H_{c//}$ are 2.8 and 1.5 kOe, respectively. This shows that the film has large perpendicular magnetic anisotropy.

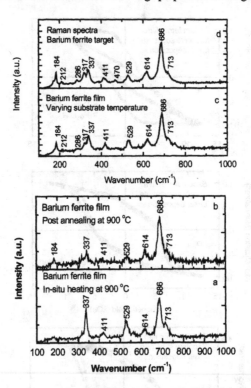

Fig. 3 The Raman spectra of the target and films prepared by laser deposition with in-situ heating, post annealing and varying substrate temperature

In contrast, for the film with post annealing as shown in Fig. 2(b), the magnetization hysteresis loops are almost the same for magnetic fields both applied in parallel and perpendicular. The perpendicular coercivity $H_{c\perp}$ and in-plane one $H_{c//}$ have the same value of 5.1 kOe. The film exhibits almost isotropic properties, which means the existence of a significant amount of in-plane c-axis orientation. However, it is not clearly shown in the x-ray spectrum [9]. Sui et al gave the two reasons to explain this phenomenon [5].

The magnetization hysteresis loops of the barium ferrite films deposited with varying substrate temperatures as shown in Fig. 2(c) shows similar for both magnetic fields applied parallel and perpendicular to the substrate surface, indicating that the films exhibit almost isotropic properties. The perpendicular and in-plane coercivities are 5.7 and 5.5 kOe for the film deposited at a temperature varying range from room temperature to 900 °C, respectively. Comparison of the magnetization hysteresis loops of the film with those films deposited with a post annealing suggests the similar magnetic properties such as isotropic properties and high coercivity [9]. While the films deposited with in-situ heating has large magnetic anisotropy and lower coercivity [14].

Figure 3 shows the Raman spectra of the target and films prepared by laser deposition with in-situ heating, post annealing and varying substrate temperature. Based on D_{6h} symmetry of barium ferrite crystals, 42 Raman-active modes ($11\ A_{1g} + 14\ E_{1g} + 17\ E_{2g}$) are expected [15]. The A_{1g} modes at 317 cm^{-1}, 411 cm^{-1}, 470 cm^{-1}, 614 cm^{-1}, 686 cm^{-1} and 713 cm^{-1}, E_{1g} modes at 184 cm^{-1}, 212 cm^{-1} and 286 cm^{-1} and E_{2g} modes at 337 cm^{-1} and 529 cm^{-1} are observed in the Raman spectra. By comparison with Raman modes between the target and the deposited films, it was found that the Raman modes of the film are the same with that of the target. Therefore, we can conclude that compositions of the film are close to that of the target ($BaFe_{12}O_{19}$). We also found that E_{1g} modes are exhibited in Figs. 3(b) to 3(d) for the target and films with post annealing and varying substrate temperature. According to Raman spectra analysis in Ref. 15, the measurement of phonons with E_{1g} symmetry is indicative of a polycrystal or at least a disoriented crystal structures. For the films with post annealing and varying substrate temperature, the E_{1g} modes can be seen. This indicates that the films contain polycrystalline grains or at least disoriented crystals, which means that there is a significant amount of in-plane c-axis orientation besides c-axis orientation normal to the plane. However, for the film with in-situ heating, no E_{1g} modes can be observed, which indicates that the film has a good orientation. Therefore, the VSM results and discussions are further consolidated by the Raman analyses.

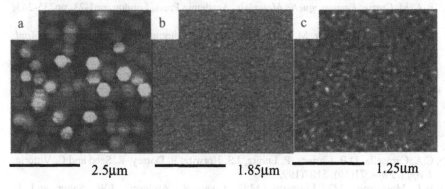

a   b   c

2.5µm   1.85µm   1.25um

Fig. 4 AFM profiles of the barium ferrite films prepared by laser deposition with (a) in-situ heating, (b) post annealing and (c) varying substrate temperature

AFM profiles of the barium ferrite films are shown in Fig. 4. It is observed that the grains in the film with in-situ heating as shown in Fig. 4(a) have good crystallinity with hexagonal symmetry. This indicates that the film is grown with c-axis normal to the film plane. While the grains in the films with post annealing and varying substrate temperature as shown in Figs. 4(b) and 4(c) show both circular and elongated shapes. Since the c-axis is normal to the long axis of the grain for barium ferrite structure [5], the c-axis is in the plane for that elongated shape grain. This indicates the existence of a significant amount of in-plane c-axis orientation. From Fig. 4, it is observed that the grain size in the film with in-situ heating is much larger than that with post annealing or varying substrate temperature. Since the coercivity increases with decreasing grain size [3], the films with post annealing and varying substrate temperature have larger coercivity than that of the film with in-situ heating, which is shown in Fig. 2.

CONCLUSION

In summary, the properties of $BaFe_{12}O_{19}$ films prepared on (001) sapphire substrates by laser deposition with in-situ heating, post annealing and varying substrate temperature have been studied. The film with in-situ heating exhibits a preferential c-axis orientation normal to the film plane and has large perpendicular magnetic anisotropy. The grains in the film have good crystallinity with hexagonal symmetry. The perpendicular coercivity and in-plane one are 2.8 and 1.5 kOe, respectively. While the film with post annealing exhibits both c-axis orientation normal to the film plane and in-plane c-axis orientation and has almost isotropic properties. The grains in the film show both circular and elongated shapes. The perpendicular coercivity and in-plane one are all 5.1 kOe. The perpendicular and in-plane coercivities are 5.7 and 5.5 kOe for the film deposited with a varying substrate temperature at 900 °C. The magnetic properties, grain shape and crystalline orientation of the film with a varying substrate temperature during film growth are close to those of the film deposited with a post annealing and different to those of the film deposited with in-situ heating.

REFERENCES

1. Okamoto, I. Kaitsu, H. Akimoto, K. Sato, E.N. Abarra and M. Shinohara, *IEEE Trans. Magn.,* 35, 2655(1999).
2. R.A. McCurrie, *Ferromagnetic Materials*, Academic Press, London, pp1-23, pp235-243, 1994.
3. T.L. Hylton, M.A. Parker, M. Ullah, K.R. Coffey, R. Umphress and J.K.Howard, *J. Appl. Phys.* 75(10), 5960(1994).
4. K. Sin, J.M. Sivertsen, J.H. Judy, *J. Appl. Phys.* 75(10), 5972(1994).
5. X. Y. Sui, M.H. Kryder, B.Y. Wong and D.E. Laughlin, *IEEE Trans. Magn.,* 29(6), 3751(1993).
6. T.L. Hylton, M.A. Parker and J.K. Howard, *Appl. Phys. Lett.,* 61(7), 867(1992).
7. Y. Hoshi, Y. Kubota and M. Naoe, *IEEE Trans. Magn.,* 31(6), 2782(1995).
8. A. Morisako, M. Matsumoto and M. Naoe, *J. Magn. Magn. Mater.,* 193, 110(1999).
9. Y.F. Lu, W.D. Song, Appl. Phys. Lett., 76(4), 490(2000).
10. C.A. Carosella, D.B. Chrisey, P. Lubitz, J.S. Horwitz, P. Dorsey, R. Seed and C. Vittoria, *J. Appl. Phys.* 71(10), 5107(1992).
11. H.J. Masterson, J.G. Lunney, J.M.D. Coey, R. Atkinson, I.W. Salter and P. Papakonstantinou, *J. Appl. Phys.* 73(8), 3917(1993).
12. R.G. Welch, T.J. Jackson and S.B. Palmer, *IEEE Trans. Magn.,* 31, 2752(1995).
13. A. Lisfi, J.C. Lodder, P.de Haan, M.A. Smithers and F.J.G. Roesthuis, *IEEE Trans. Magn.,* 34, 1654(1998).
14. W.D. Song, Y.F. Lu, W.J. Wang, C.K. Ong, J.P. Wang and T.C. Chong, *IEEE. Trans. Magn.* 35(5), 3013(1999).
15. S. Pignard, H. Vincent, J.P. Senateur and G. Lucazeau, *Appl. Phys. Lett.,* 73(9), 1194(1998).

Mat. Res. Soc. Symp. Vol. 617 © 2000 Materials Research Society

ROOM TEMPERATURE GROWTH OF INDIUM TIN OXIDE FILMS BY ULTRAVIOLET-ASSISTED PULSED LASER DEPOSITION

V. CRACIUN*, D. CRACIUN**, Z. CHEN*, J. HWANG***, R.K. SINGH*
*Materials Science & Engineering, University of Florida, Gainesville, FL 32611
**National Institute for Laser, Plasma and Radiation Physics, Bucharest, Romania
***Physics Department, University of Florida, Gainesville, FL 32611

ABSTRACT

The characteristics of indium tin oxide (ITO) films grown at room temperature on (100) Si and Corning glass substrates by an in situ ultraviolet-assisted pulsed laser deposition (UVPLD) technique have been investigated. The most important parameter, which influenced the optical and electrical properties of the grown films, was the oxygen pressure. For oxygen pressure below 1 mtorr, films were metallic, with very low optical transmittance and rather high resistivity values. The resistivity value decreased when using higher oxygen pressures while the optical transmittance increased. The optimum oxygen pressure was found to be around 10 mtorr. For higher oxygen pressures, the optical transmittance was better but a rapid degradation of the electrical conductivity was noticed. X-ray photoelectron spectroscopy investigations showed that ITO films grown at 10 mtorr oxygen are fully oxidized. All of the grown films were amorphous regardless of the oxygen pressure used.

INTRODUCTION

Indium tin oxide (ITO) thin films are widely used for optoelectronic devices as they combine a good electrical conductivity with high transparency in the visible range. There are a number of interesting applications such as anode contact in organic light-emitting diodes [1, 2] or coating of flexible polymer substrates for ultralight mobile display panels [3] where the use of a low processing temperature is very important. The use of the pulsed laser deposition (PLD) technique, which has several important advantages [4], has allowed the growth of good quality indium tin oxide (ITO) thin films at relatively low temperatures and even room temperature [5-8]. Other techniques such as synchrotron radiation ablation [9] or plasma-ion assisted evaporation [10] were also employed to deposit ITO films at room temperature. We investigated the use of an in situ ultraviolet-assisted PLD technique (UVPLD) for the growth of ITO films at room temperature. The UV source photodissociates molecular oxygen and provides ozone and atomic oxygen during the growth [11]. These more reactive gases have been shown to promote the crystalline growth at lower temperatures than those normally used during conventional PLD [12]. Moreover, UV+ozone is known to be an effective way to clean organic contaminants from the substrate [13, 14], a fact that can also improve the quality of the deposited layers.

EXPERIMENT

The PLD system employed is presented elsewhere in much more detail [15, 16] and it is only briefly described here. An excimer laser (KrF, λ=248 nm, laser fluence ~2 J/cm^2, repetition rate 5 Hz) was used to ablate ITO targets (99.99% purity). The oxygen pressure was varied from 0.1 mtorr to 50 mtorr. A vacuum compatible, low pressure Hg lamp, which allows for in-situ UV irradiation of the substrate during both the laser ablation-growth process, was fitted to the PLD system. Films were deposited onto (100) Si wafers and Corning glass substrates that were placed at 10.5 cm in front of the target.

The crystalline structure of the grown films was investigated by x-ray diffraction (XRD) and transmission electron microscopy (TEM). The chemical composition was determined by x-ray photoelectron spectroscopy (XPS, Perkin Elmer 5100, Mg Kα radiation) and Auger electron spectroscopy (AES). The optical properties of films grown on Si substrates were investigated by spectroscopic ellipsometry (VASE, Woollam Co.) at 70° while those of film deposited on Corning glass by optical spectrophotometry. The sheet resistance was measured by a four point probe method and the surface morphology by atomic force microscopy (AFM).

RESULTS

XRD and TEM investigations showed that all films deposited, regardless of the oxygen pressure used, were amorphous. Crystalline films were deposited by UVPLD at a substrate temperature of only 100 °C. The effect of the oxygen pressure on the resistivity of the grown ITO films is shown in Fig. 1. Values measured for several films, which were grown by conventional PLD for comparison reasons, are also shown in Fig. 1. As one can note, there is a continuous improvement of the electrical conductivity with the increase of the oxygen pressure up to a value of 10 mtorr. A further increase of the oxygen pressure resulted in a steep increase of the electrical resistivity. A similar dependence has been already reported for ITO films grown by PLD [7, 8]. It is worth noting that by using the UVPLD technique it was possible to obtain resistivity values below 4×10^{-4} Ω.cm.

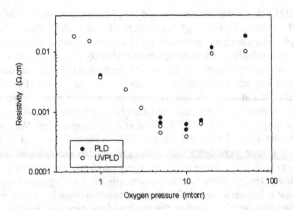

Figure 1. Resistivity values versus oxygen pressure for ITO films grown by PLD and UVPLD.

The optical transmittance of a typical ITO film grown under the optimum oxygen pressure, i. e. 10 mtorr, is displayed in Fig. 2. One can note that the film exhibited a transmission almost as good as the Corning glass substrate. The refractive index and extinction coefficient values estimated from the transmittance curve are shown in Table I. The optical transmittance of a film grown under 2 mtorr oxygen is also shown. It is obvious that the effect of oxygen pressure on the optical transmittance is very important.

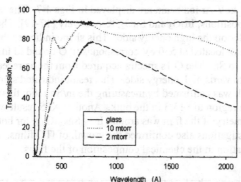

Figure 2. Optical transmittance of ITO thin films deposited by UVPLD on Corning glass.

Table I. Optical parameters of the ITO film showed in Fig. 2.

Thickness	Wavelength	Refractive index	Extinction coefficient
200 nm	422 nm	2.1	0.018
	500 nm	2.0	0.011
	730 nm	1.9	0.003
	1120 nm	1.5	0.016

The optical properties of ITO films deposited on Si were investigated by VASE. In Fig. 3 the refractive index and extinction coefficient values of films deposited by PLD (328 nm thick) and UVPLD (317 nm thick) under 10 mtorr oxygen are shown. One can note that apart from the region of high absorption (270-310 nm) the refractive index values were quite close to the reference ITO values. Also, the extinction coefficient values were very small, confirming the transparency of these films determined by spectrophotometry.

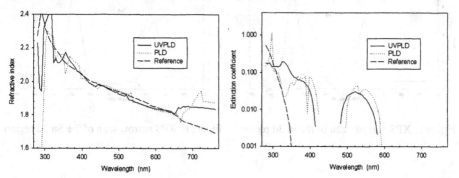

Figure 3. Refractive index and extinction coefficient values of ITO films grown under 10 mtorr.

XPS investigations were performed on samples grown on Si under 5 and 10 mtorr of oxygen by conventional PLD and UVPLD. Typical examples of a survey and high resolution scans for In $3d_{5/2}$, Sn $3d_{5/2}$, and O 1s spectra acquired from a sample that was in situ sputtered cleaned for 3 min by a 2 keV Ar ion beam are displayed in Figs 4-7. The In and Sn peaks correspond to oxidized positions, In_2O_3 and SnO_2, respectively [17-19]. The oxygen peak exhibited a small shoulder on the high energy side. This shape indicates the presence of two oxidation states. The peak located at 530 eV corresponds to O bound to In and that at 531.0 eV corresponds to O bound to Sn. The O 1s spectra acquired from the as-received surface exhibited a much larger shoulder towards high energy side. This result indicated some Sn segregation in the surface region, which was confirmed by measuring the In/Sn ratio: this increased from around 10 in the surface region to ~13.3 in the bulk. Another XPS result worth mentioning is that the overall stoichiometry of the film was around $In_{0.44}Sn_{0.03}O_{0.53}$ for both PLD and UVPLD grown films. AES investigations also confirmed the growth of ITO films. It also showed that there was no lateral variation in the chemical composition of the films.

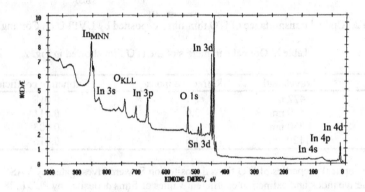

Figure 4. XPS survey spectra of an ITO film grown by UVPLD under 10 mtorr O_2; the sample was sputtered clean with 2.0 keV Ar ions for 3 min; take-off angle 45°.

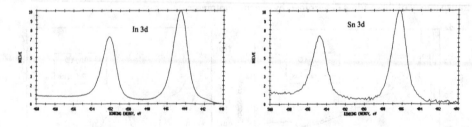

Figure 5. XPS narrow scan of the In 3d region.   Figure 6. XPS narrow scan of the Sn 3d region.

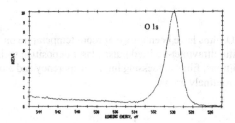

Figure 7. XPS narrow scan of the O 1s region

The surface morphology was investigated by AFM. A typical image is displayed in Fig. 8. The films were quite smooth, with an average root-mean-square (RMS) of around 0.3 nm. This is a very good value, comparable to other reported results [8, 17].

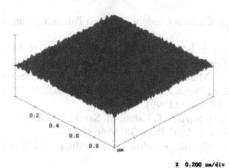

X 0.200 um/div
Z 10.000 nm/div

Figure 8. AFM image (1 μm x 1μm) of the ITO film grown on Si.

It has been found that UVPLD grown films exhibited slightly better electrical and optical properties than PLD grown films under identical conditions. However, neither XPS nor TEM investigations could find substantial differences in the microstructure or composition between these films. UV radiation and ozone help to better clean the substrate. According to XPS results there was always a slightly lower C 1s peak due to contamination on UVPLD grown samples. The ozone and atomic oxygen formed by photodissociation are more reactive than molecular oxygen and can oxidize metals even at room temperature. Also, because In and Sn can react with oxygen during PLD even in gas phase [5], it is very likely they will react even more with ozone. Also, the surface mobility of adatoms is increased by UV irradiation [20]. It was found that under identical conditions the PLD grown samples were a little bit thicker than UVPLD grown ones, implying that the desorption of loosely bound adatoms was also increased during UV irradiation. All these factors could contribute to the observed better characteristics of UVPLD grown ITO films.

CONCLUSIONS

Good quality ITO films have been grown at room temperature on Si and Corning glass substrates using an in situ ultraviolet-assisted pulsed laser deposition technique. Under optimum oxygen deposition conditions, films possessing high transparency and a resistivity lower than 4×10^{-4} Ω.cm have been routinely grown.

REFERENCES

1. H. Kim, A. Pique, J. S. Horwitz, H. Mattoussi, H. Murata, Z. H. Kafafi, and D. B. Chrisey, Appl. Phys. Lett. **74**, 3444 (1999).
2. C. W. Tang and S. A. Van Slyke, Appl. Phys. Lett. **51**, 913 (1987).
3. F. Matsumoto, Asia Display'95, p. 31 (1995).
4. D. B. Chrisey and G. K. Hubler, Pulsed Laser Deposition of Thin Films, Wiley, New York, 1994.
5. R. Teghil, V. Marotta, A. Giardini Guidoni, T. M. Di Palma, C. Flamini, Appl. Surf. Sci. **138-139**, 522 (1999).
6. F. O. Adurodija, H. Izumi, T. Ishira, H. Yoshioka, K. Yamada, H. Matsui, and M. Motoyama, Thin Solid Films **350**, 79 (1999).
7. Y. Yamada, N. Suzuki, T. Makino, and T. Yoshida, J. Vac. Sci. Technol. A **18**, 83 (2000).
8. H. Kim, C. M. Gilmore, A. Pique, J. S. Horwitz, H. Mattoussi, H. Murata, Z. H. Kafafi, and D. B. Chrisey, J. Appl. Phys. **86**, 6451 (1999).
9. Y. Akagi, K. Hanamoto, H. Suzuki, T. Katoh, M. Sasaki, S. Imai, M. Tsudagawa, Y. Nakayama, and H. Miki, Jpn. J. Appl. Phys. **38**, 6846 (1999).
10. S. Laux, N. Kaiser, A. Zoller, R. Gotzelmann, H. Lauth, and H. Bernitzki, Thin Solid Films **335**,1 (1998).
11. D. L. Baulch, R. A. Cox, R. F. Hampson, Jr.,J. A. Kerr, J. Troe, and R.T. Watson, J. Phys. Chem. Ref. Data **9**, 295 (1980).
12. C. E. Otis, A. Gupta, and B. Braren, Appl. Phys. Lett. **62**, 102 (1993).
13. S. K. So, W. K. Choi, C. H. Cheng, L. M. Leung, and C. F. Kwong, Appl. Phys. A **68**, 447 (1999).
14. K. Sugiyama, H. Ishii, Y. Ouchi, and K. Seki, J. Appl. Phys. **87**, 295 (2000).
15. V. Craciun and R.K. Singh, Electrochemical and Solid-State Lett. **2**, 446 (1999).
16. V. Craciun, R.K. Singh, J. Perriere, J. Spear, and D. Craciun, Appl. Phys. A **69** [Supl.], S531 (1999).
17. S-S. Kim, S-Y. Choi, C-G. Park, H-W. Jin, Thin Solid Films **347**, 155 (1999).
18. D. Rhode, H. Kersten, C. Eggs, and R. Hippler, Thin Solid Films **305**, 164 (1997).
19. F. Zhu, C. H. A. Huan, K. Zhang, and A. T. S. Wee, Thin Solid Films **359**, 244 (2000).
20. H. Wengenmair, J. W. Gerlach, U. Preckwinkel, B. Stritzker, and B. Rauschenbach, Appl. Surf. Sci. **99**, 313 (1996).

Mat. Res. Soc. Symp. Vol. 617 © 2000 Materials Research Society

Oxygen Content and Crystallinity Effects in Pulsed Laser Deposited Lanthanum Manganite Thin Films

Srinivas V. Pietambaram, D. Kumar, Rajiv K. Singh and C. B. Lee[1]
Department of Materials Science and Engineering, University of Florida, Gainesville, Florida 32611-6400, U.S.A.
[1]Department of Electrical Engineering, North Carolina A & T University, Greensboro, North Carolina 27411, U.S.A.

ABSTRACT

Eventhough colossal magnetoresistance in Lanthanum calcium manganite (LCMO) thin films was known for a long time, the effect of oxygen content and crystallinity on the properties of these films is not clearly understood. It is in this context that we have performed a systematic study of these effects by annealing the films in various ambients. A series of LCMO thin films have been grown *in situ* on (100) LaAlO₃ substrates using a pulsed laser deposition technique under identical conditions. Microstructural characterization carried out on these films has shown that the films are smooth, single phase and highly textured. The films were subjected to the following post deposition treatments: (i) annealing in oxygen at 900°C for 4 hrs, (ii) annealing in argon at 900°C for 4 hrs, (iii) annealing in oxygen at 500°C for 12 hrs, (iv) annealing in argon for 12 hrs and (v) annealing in vacuum at 850°C for half-an-hour. As deposited LCMO films show a transition temperature of 260 K and a magnetoresistance ratio (defined as [R(0)-R(H)/R(H)]) of 190% at 260 K in 5T magnetic field. The samples subjected to 500°C oxygen and Ar anneal have shown no change in the transition temperature and MR ratio. The films subjected to a 900°C annealing in Ar ambient have shown marginal improvement in transition temperature but a drastic improvement in the MR ratio (525%). 900°C oxygen annealed films have shown an improvement in the transition temperature (290 K) and MR ratio (225%) over as deposited films. Vacuum annealed samples have shown deteriorated properties. These results indicate that the metal-insulator transition is related to the oxygen content of the films while the MR ratio is related to the domain size.

INTRODUCTION

Colossal magnetoresistance (CMR) was first observed in single crystal La₀.₆Pb₀.₄MnO₃ by Searle and Wang [1]. Recent interest in these materials was sparked off by the observation of large magnetoresistance in epitaxial thin films near room temperature [2]. Thin films with large room temperature magnetoresistance open up new possibilities for applications in diverse areas of technology such as magnetic random access memories and read heads for hard disk drives. The effects of oxygen in manganite film have been a focus since the discovery of extraordinary magnetotransport in this kind of materials. There is no quantitative relation between the oxygen content and the magnetic and transport properties in thin films to date due to the difficulty in the determination and control of oxygen content. Substrate temperature, oxygen partial pressure, and deposition rate could have an effect on the oxygen content of the film severely [3-5]. Vacuum annealing can produce a deficiency in oxygen [6] while excess oxygen can be produced by annealing in oxygen [7,8]. All these indicate that oxygen content can be flexible in manganese oxides. In the early studies of CMR materials, a post deposition anneal in oxygen at high temperatures was critical for achieving large magnetoresistance [2,3]. However, it was not known whether the improvement was due to grain refinement through grain growth and enhancement in the crystallinity of the films or due to oxygen incorporation. Improvement in the crystallinity and grain growth enhance the properties of the oxide thin films [9,10]. In this paper, we report our results of systematic post deposition heat treatments performed to deconvolute the effects of oxygen content and grain growth in Lanthanum manganite thin films.

EXPERIMENT

Bulk $La_{0.7}Ca_{0.3}MnO_3$ (LCMO) was prepared by a ceramic method. The required quantities of respective oxide or carbonate powders were mixed and sintered at 1400°C for 24 hours. Six LCMO films with a thickness of 1500A were grown *in situ* on (100) $LaAlO_3$ substrates using a pulsed laser ablation system. To eliminate other external effects, all the substrates were placed side by side on sample holder. All the films were characterized to see if the properties were identical before performing any post deposition heat treatments. A detailed description of the deposition system is mentioned elsewhere [11]. In brief, a 248 nm KrF pulsed laser with 5 Hz repetition rate and 1.6 J/cm² energy density was used. A substrate temperature of 700°C and oxygen pressure of 250 mTorr were used during the deposition of the films. Following the deposition, the films were cooled down to room temperature at a rate of 10°C/min in 400 Torr of oxygen. After the initial characterization to check the identical properties, the films were subjected to the following post deposition anneals – (i) annealing in oxygen at 900°C for 4 hrs, (ii) annealing in argon at 900°C for 4 hrs, (iii) annealing in oxygen at 500°C for 12 hrs, (iv) annealing in argon for 12 hrs and (v) annealing in vacuum at 850°C for half-an-hour.

The films deposited were characterized using scanning electron microscopy (SEM), energy dispersive x-ray analysis (EDX), and x-ray diffraction (XRD) measurements. The temperature dependence of resistance of the films was examined in zero and applied field using four-probe technique and the quantum design superconducting quantum interference device (SQUID) magnetometer. Both the transport current and applied field were in the film plane. The MR ratio, $\Delta R/R(H)$, was calculated using $\Delta R/R(H) = [R(H)-R(0)]/R(H)$, where R(H) and R(0) are resistances in applied and zero field.

RESULTS AND DISCUSSION

XRD patterns of the films subjected to various anneals are shown in Fig. 1. From the figure it is clear that all the films have single phase with (00*l*) peaks with *l* = 1 and 2. The presence of only sharp (00*l*) peaks indicate the highly textured growth of all the films on (100)

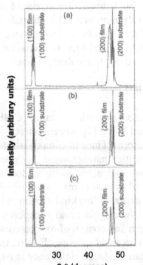

Fig.1. XRD patterns of LCMO films, (a) no anneal, (b) 900°C oxygen anneal, and (c) 900°C argon anneal

LaAlO₃ substrate. The lattice parameters for the films with no anneal, 900C oxygen anneal and 900C argon anneal were found to be 3.8632, 3.8337, and 3.8444 A respectively. XRD patterns of 500C oxygen and argon annealed films were similar to that of as-deposited films. The films annealed in oxygen and argon show a decrease in the full width at half maximum (0.1°) compared to the as-deposited film (0.25°) indicating these films are more crystalline.

The variation of electrical resistance in zero and applied field (5T) as a function of temperature for the films subjected to various anneals are shown in Fig. 2. All the films were grown under identical conditions so that film thickness, oxygen contents of all the films and other external effects (before anneals) could be kept identical. It is important to keep these parameters identical in order to reveal the effects of the magnetotransport properties in these films. According to the

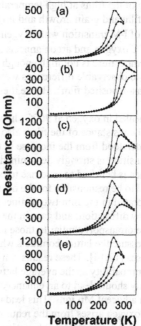

Fig.2. Variation of electrical resistance with temperature for the following films (a) no anneal, (b) 900⁰C oxygen anneal, (c) 900⁰C argon anneal, (d) 500⁰C oxygen anneal and (e) 500⁰C argon anneal (filled circle (0T), filled diamond (2T) and filled triangle (5T))

variation of resistance shown in Fig. 2., all the films (except vacuum annealed one), have similar qualitative magnetotransport behavior. That is, all the films undergo an insulator-to-metal (I-M) transition as the temperature is lowered down and the resistance of all the films is suppressed significantly with the application of magnetic field. The suppression in film resistance in each case is maximum near the resistivity peak in zero field as observed frequently by others in several manganite systems [1-3]. The MR ratios of the films were calculated using the data in Fig.2. The MR ratios obtained are plotted in Fig.3 as a function of temperature at 5T.

The as-deposited films show an I-M transition at 260 K and a MR ratio of 190% in 5T. The films subjected to 500°C oxygen anneal and 500°C argon anneal show little change in I-M transition and MR ratio. This shows that 500°C is not a sufficiently high enough temperature to affect the oxygen content, grain growth and crystallinity. The films subjected to 900°C oxygen anneal show an improvement in transition temperature (290 K) but a marginal enhancement in the MR ratio (225%). The films subjected to 900°C argon anneal show slight increase in the transition temperature (270 K) and significant enhancement in the MR ratio (525%). The films subjected to vacuum anneal have shown deteriorated properties (insulating down to 10 K).

Annealing of the films at high temperature leads to two simultaneous effects – removal of oxygen from the film and grain growth and improvement in crystallinity. Ju et al. [6] established a linear decrease of I-M transition with oxygen deficiency for the bulk $La_{0.67}Ba_{0.33}MnO_{3-\delta}$. Films subjected to 500°C oxygen and argon anneals show little effect on the transition temperature or MR ratio. This temperature is not high enough for either oxygen incorporation or to cause grain growth. Further, no observable differences were found from the XRD patterns of these films as compared to the as-deposited films. Hence these anneals have no effect on the properties of the film.

Films annealed in oxygen at 900C show an increase in the transition temperature from 260K to 290K. The resistance of the films also decreased compared to as deposited films. These effects can be understood from the increase in the oxygen content of the films. The ferromagnetic transition is strongly determined by the number of Mn^{4+} ions. The mixed Mn^{3+}/Mn^{4+} valence is believed to give rise to both ferromagnetism and metallic behavior in LCMO films, and to be responsible for the occurrence of colossal magnetoresistance [12]. As oxygen is incorporated in the film two distinct reactions occur: contraction of the lattice, as evidenced by x-ray diffraction; and gain in the O^{2-} ions. Gain of oxygen ions should lead to changes in magnetotransport similar to those resulting from the application of external pressure. Under applied pressure the lattice contracts while transition temperature (T_c) increases and resistance decreases [13,14]. These results can be explained by an enhancement in the Mn-Mn electron transfer probability as the average lattice spacing is decreased. An increase in the transfer probability should lead to an enhanced ferromagnetic correlations and increased carrier mobility between adjacent Mn ions. This leads to a higher T_c and lower resistance. The second effect relating to oxygen arises from the requirement of charge neutrality within each unit cell. The chemical formula for LCMO can be written as $La_{1-x}^{3+}Ca_x^{2+}Mn_{1-x+2\delta}^{3+}Mn_{x-2\delta}^{4+}O_{3-\delta}^{2-}$. Therefore, each oxygen incorporated into LCMO should lead to a conversion of Mn^{3+} ions to two Mn^{4+} ions. The carriers in LCMO are holes whose concentration is proportional to the Mn^{4+} concentration [15]. Therefore, the incorporation of oxygen should increase Mn^{4+} concentration which leads to an increase in the carrier concentration and hence decrease in the resistance.

However, annealing at high temperatures also causes grain growth which affects the domain size. These effects can be isolated by annealing the films at elevated temperature in an ambient other than oxygen such as argon. Films subjected to argon annealing at 900C show marginal increase in the transition temperature. The resistance of the films is higher than that of the oxygen annealed films. The increase in the resistance of the films can be attributed to the loss of the oxygen from these films. The interesting

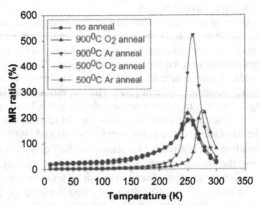

Fig.3. Variation of MR ratio with temperature for the films

feature of high temperature argon annealed films is significantly high MR ratio observed in these films. This may be attributed to the increase in the domain size. Near the domain-wall boundaries, the pairs of spins of Mn^{3+} and Mn^{4+} may not be parallel. As a result the electron transfer between pairs of Mn^{3+} and Mn^{4+} ions across the domain wall is difficult and the resistance is high. Increase in the domain size reduces the number of domain-wall boundaries in a specified area. As a result, smaller fields are necessary to align these domains, hence the suppression of resistance is much higher for a give magnetic field and as such higher MR ratios. The marginal improvement in the transition temperature may also be related to an increased domain size due to the smaller amount of domain-wall boundaries and reduced resistance for the ferromagnetic alignment. The lattice parameter of argon annealed films is higher than that of oxygen annealed films which indicates that the films have less oxygen. However, the decrease in the lattice parameter in argon annealed films compared to as-deposited films is not clearly understood. It seems that an increase in the domain size causes a decrease in the lattice parameter.

Vacuum annealed samples undergo the most severe heat treatment. The loss of oxygen from the film leads to the conversion of two Mn^{4+} ions to Mn^{3+} ions. Sun et al. [16] suggested the presence of critical oxygen content for the occurrence of metal to insulator transition (MI) in the LCMO films. According to Cheong et al. [17], MI transition will disappear in the compound $La_{1-x}Ca_xMnO_3$ with a Mn^{4+}/Mn^{3+} ratio less than 0.17. In the vacuum annealed samples, there may be a depletion of oxygen to the extent in which Mn^{4+}/Mn^{3+} ratio is less than 0.17. Hence these films show insulating behavior down to 10 K.

CONCLUSIONS

In summary, we have deposited LCMO films under identical conditions and subjected them to systematic anneals in various ambients. Low temperature (500°C) oxygen and argon anneals show little effect on the transition temperature or MR ratio. High temperature (900°C) oxygen anneal shows a significant improvement in the transition temperature while high temperature (900°C) argon anneal shows a substantial increase in the MR ratio. The transition temperature is related to the oxygen content while the MR ratio is affected by the domain size.

REFERENCES

1. C. W. Searle and S. T. Wang, *Can. J. Phys.* **48**, 2023 (1970).
2. R. Von Helmolt, J. Wecker, B. Holzapfel, L. Schultz, and K. Samwer, *Phys. Rev. Lett.* **71**, 2331 (1993).
3. S. Jin, T. H. Tiefel, M. McCormack, R.A. Fastnatch, R. Ramesh, and L. H. Chen, *Science* **264**, 413 (1994).
4. G. C. Xiong, Qi Li, H. L. Ju, R. L. Greene, and T. Venkatesan, *Appl. Phys. Lett.* **66**, 1689 (1995).
5. Z. Wei, W. Boyd, M. Elliot, and W. Herrenden-Harkerand, *Appl. Phys. Lett.* **69**, 3926 (1996).
6. H. L. Ju, J. Gopalakrishnan, J. L. Peng, Qi Li, G. C. Xiong, T. Venkatesan and R. L. Greene, *Phys. Rev. B* **51**, 6143
7. J. F. Mitchell, D. D. Argyriou, C. D. Potter, D. G. Hinks, J. D. Jorgensen, and S. D. Barder, *Phys. Rev. B* **54**, 6172 (1996).

8. W. Prellier, M. Rajeswari, T. Venkatesan, and R. L. Greene, *Appl. Phys. Lett.* **75**, 1446 (1999).
9. Y. Matsumoto, j. Hombo, Y. Yamaguchi, M. Nishida, and A. Chiba, *Appl. Phys. Lett.* **56**, 1585 (1990).
10. R. K. Singh, D. Bhattacharya, P. Tiwari, J. Narayan, and C. B. Lee, *Appl. Phys. Lett.* **60**, 255 (1992).
11. Srinivas V. Pietambaram, D. Kumar, Rajiv K. Singh, and C. B. Lee, *Phys. Rev. B* **58**, 8182 (1998).
12. C. Zener, *Phys. Rev.* **82**, 403 (1951).
13. H. W. Hwang, T. T. M. Palstra, S.-W. Cheong, and B. Batlogg, *Phys. Rev B* **52**, 15046 (1995).
14. Y. Moritomo, A. Asamitsu, and Y. Tokura, *Phys. Rev. B* 51, 16491 (1995).
15. A. Urushibara, Y. Moritomo, T. Arima, A. Asamitsu, G. Kido, and Y. Tokura, *Phys. Rev. B* **51**, 14103 (1995).
16. P. Schiffer, A. P. Ramirez, W. Bao, and S.-W. Cheong, *Phys. Rev. Lett.* **75**, 3336 (1995).
17. J. R. Sun, C. F. Yeung, K. Zhao, L. Z. Zhou, C. H. Leung, H. K. Wong, and B. G. Shen, *Appl. Phys. Lett.* **76**, 1164 (2000).

Mat. Res. Soc. Symp. Vol. 617 © 2000 Materials Research Society

RAMAN STUDIES OF LiMn$_2$O$_4$ FILMS GROWN BY LASER ABLATION

M.A. CAMACHO-LOPEZ[1,2], L. ESCOBAR-ALARCON[3], E. HARO-PONIATOWSKI[2], C. JULIEN[1]
[1]LMDH, UMR 7603, Université Pierre et Marie Curie, 4 place Jussieu, case 86
75252 Paris cedex 05, France
[2]Laboratorio de Optica Cuantica, Universidad Autónoma Metropolitana Iztapalapa
Apdo. Postal 55-534, México DF 09340, México
[3]Departamento de Física, Instituto Nacional de Investigaciones Nucleares
Apdo. Postal 18-1027, México DF 11801, México

ABSTRACT

Polycrystalline thin films of lithium manganese oxide have been grown using the pulsed-laser deposition (PLD) technique. Films of LiMn$_2$O$_4$ were deposited onto Si substrates heated at temperature lower than 300°C from a sintered composite target (LiMn$_2$O$_4$+Li$_2$O) irradiated with a Nd:YAG laser. The structural characterizations of these films have been carried out by Raman scattering spectroscopy which probe the local environment of cations in the LiMn$_2$O$_4$ framework. Raman spectra of PLD LiMn$_2$O$_4$ films have been investigated as a function of various growth conditions, i.e. substrate temperature, partial oxygen pressure in the deposition chamber, and target composition.

INTRODUCTION

Lithiated transition-metal oxides (LTMOs) such as LiCoO$_2$, LiNiO$_2$, and LiMn$_2$O$_4$ have received significant attention due to their industrial applications especially in rechargeable lithium-ion batteries [1]. These materials are applied on the cathode side where Li is respectively extracted and stored during the charge-discharge cycle of the battery. It has been reported that the spinel structure LiMn$_2$O$_4$ exhibits a specific capacity about 120 mAh/g, where composite sample electrodes were used. The classical insertion/deinsertion reactions in a Li//LiMn$_2$O$_4$ cell normally lie in the voltage range 3.6-4.4 V [2].

Preparation of LiMn$_2$O$_4$ in thin-film form may have advantages from a point of view of fundamental studies (because it is a binder-free material with a well-defined interfacial area) and of the emerging field of microbatteries as well. Thin films of LTMOs have been synthesized by a variety of techniques including sputtering, spray deposition, reaction of metals, and pulsed laser deposition [3-13]. In the fabrication of LiMn$_2$O$_4$ thin films, formation of the spinel structure is known to be crucial for obtaining a good rechargeability of the cells. Various aspects of LiMn$_2$O$_4$ thin films prepared by RF sputtering [7], electron beam deposition [4], pyrolitic preparation [7], and pulsed-laser deposition [11-13] have been reported. Thin films of amorphous LiMn$_2$O$_4$ have been prepared onto substrates maintained at low-temperature (<150°C) by Shokoohi et al. [4] with reactive electron beam evaporation and Hwang et al. [6] with RF magnetron sputtering.

Pulsed-laser deposition (PLD) technique is a successful method in the growth of materials containing volatile components with complex stoichiometries. For this reason, it is well suited to films of LTMOs, where lithium loss due to volatilization could occur in conventional evaporation methods. Recently, Cairns et al. reported the electrochemical behavior of LiMn$_2$O$_4$ films grown by PLD onto stainless steel substrates [11-12]. The authors used a LiMn$_2$O$_4$ target irradiated at 308 nm in vacuum and under 100 mTorr oxygen atmosphere; the substrate was heated at 800°C. The growth of crystalline thin dense films without postdeposition annealing was claimed and the good electrochemical performance of PLD films was demonstrated. Recently, Morcrette et al. [13] have grown PLD LiMn$_2$O$_4$ films on various substrate materials with a good crystallinity when films were formed under 0.2 mbar at 500°C.

In the present work, we wish to show that polycrystalline LiMn$_2$O$_4$ films may be grown onto substrates maintained at low temperature ($\approx$300°C) using the pulsed-laser deposition method from a sintered LiMn$_2$O$_4$ target. The structural characterizations of these films have been carried out by various techniques, i.e., x-ray diffraction (XRD), Raman scattering (RS) spectroscopy, and electron scanning microscopy (SEM).

EXPERIMENTAL

PLD films were growth onto silicon substrates maintained at various temperatures in the range 100-300 °C. PLD targets were made of disc pellets 2 mm thick and 13 mm diameter. Tablets were prepared by mixing and crushing high purity LiMn$_2$O$_4$ and Li$_2$O powders (Cerac Products). Mixtures with various molar ratios, i.e. Li/Mn>0.5 by adding Li$_2$O to provide excess of lithium in the chamber, were pressed at 5 tons/cm^2. At this stage, the targets were very fragile, but after sintering in air at 700°C for 5 h they became quite robust. The typical substrates, i.e. Si (100) oriented wafers, were previously cleaned using HF solution. Target and substrates were placed inside a vacuum chamber with a diffusion pump yielding pressures of 5x10^{-5} Torr. The target was rotated in the range of 1-10 rotations per minute with an electric motor to avoid depletion of material at any given spot. The laser used in these experiments is the 532 nm line of a double frequency pulsed Nd:YAG using power densities close to 10^8 W/cm^2 with 10 ns pulse width at repetition rate of 10 Hz. The films were grown in a partial pressure of oxygen in the range 50-300 mTorr.

The x-ray diffraction measurements were carried out using a CuK$_\alpha$ radiation source (λ=1.5406 Å) in a Siemens D-5000 diffractometer. Raman spectroscopy measurements were performed at room temperature in air with a double monochromator (Jobin-Yvon U1000) using the 514.5 nm line of an argon-ion laser (Coherent Innova 70) at a power level of 20 mW. The laser spot size at the surface of the sample was about 100 μm^2. The signal was detected with a photomultiplier and a standard photoncounting system.

RESULTS AND DISCUSSION

Pulsed-laser deposited LiMn$_2$O$_4$ films are pin-hole free as revealed from optical microscopy and well adherent to the substrate surface. The good film integrity is favorable for electrochemical testing. Fig. 1 shows the x-ray diffraction patterns of a polycrystalline PLD LiMn$_2$O$_4$ film grown onto silicon wafers maintained at T$_s$=300°C in P(O$_2$)=100 mTorr. They are indexed using the Fd3m symmetry. When a film is grown from the target LiMn$_2$O$_4$+15% Li$_2$O, the x-ray diagram displays peaks at 2θ=16, 36 and 47°, which are attributed to the (111), (311), and (400) Bragg lines respectively [2].

Experimental Raman spectra of thin-films LiMn$_2$O$_4$ are shown at various stages of pulsed-laser deposition. Fig. 2(a-c) shows the Raman scattering spectra of LiMn$_2$O$_4$ films deposited onto silicon maintained at 300°C in oxygen partial pressure P(O$_2$)=100 mTorr as a function of the target composition. These spectra display the Raman-active mode of the silicon wafer. The strong peak centered at 521 cm^{-1} corresponding to the Si substrate is observed. The RS peak positions of PLD

LiMn$_2$O$_4$ films are in good agreement with those reported for the crystalline LiMn$_2$O$_4$ [15-17]. The experimental Raman data consist of a series of broad bands located between 300 and 700 cm^{-1}. The RS spectrum of the polycrystalline LiMn$_2$O$_4$ film grown with a lithium-rich target (curve a) is dominated by a strong and broad band at ca. 622 cm^{-1} with a shoulder at 573 cm^{-1}. A band with a medium intensity is located at ca. 499 cm^{-1}, while four bands having a weak intensity are observed at ca. 437, 374, 320 and 303 cm^{-1}. Interpretation is complicated by the appearance of a higher number of Raman bands than that predicted by factor-group theory for the perfectly spinel LiMn$_2$O$_4$ crystal [15].

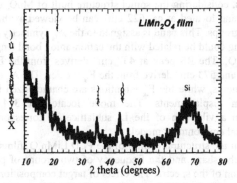

Fig. 1. X-ray diffraction patterns of pulsed-laser deposited LiMn$_2$O$_4$ films grown at T$_s$=300°C and P(O$_2$)=100 mTorr (LIMN5) from target LiMn$_2$O$_4$+15% Li$_2$O.

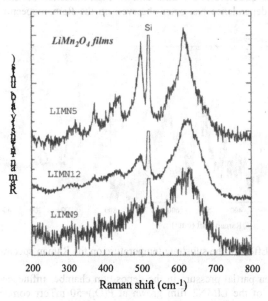

Fig. 2. Raman scattering spectra of LiMn$_2$O$_4$ films deposited onto silicon maintained at 300°C in oxygen partial pressure P(O$_2$)=100 mTorr as a function of the target composition. (a) LIMN5

grown from LiMn$_2$O$_4$+15% Li$_2$O, (b) LIMN12 grown from LiMn$_2$O$_4$+10% Li$_2$O, and (c) LIMN grown from LiMn$_2$O$_4$+5% Li$_2$O.

The cubic spinel possesses the prototype symmetry Fd3m with the standard notatic Li$_{8a}$[Mn$_2$]$_{16d}$O$_4$, where the manganese cations reside on the octahedral 16d sites, the oxygen anio on the 32e sites, and the lithium ions occupy the tetrahedral 8a sites [20]. According to the resul of the theoretical factor-group analysis, five modes are active (A$_{1g}$+E$_g$+3F$_{2g}$) in the Raman spectru of the spinel LiMn$_2$O$_4$ crystal [15]. It is also convenient to analyze the RS spectrum in terms c localized vibrations, considering the spinel structure built of MnO$_6$ octahedra and LiO$_4$ tetrahed [16]. The Raman band located at ca. 622 cm^{-1} can be viewed as the symmetric Mn-O stretchin vibration of MnO$_6$ groups. This band is assigned to the A$_{1g}$ symmetry in the O$_h^7$ spectroscopic spac group. Its broadening could be related with the cation-anion bond lengths and polyhedral distortio occurring in LiMn$_2$O$_4$. The RS peak at 437 cm^{-1} derives from the E$_g$ species, whereas the peal located at 374, 499 and 573 cm^{-1} derive from the F$_{2g}$ species. The A$_{1g}$ and E$_g$ vibrations involve on oxide ion displacements, while the F$_{2g}$ vibrations are characterized by large oxygen motions an very small lithium displacements. The mode located at 374 cm^{-1} (F$_{2g}$ species) deriv predominantly from a vibration of the Li sublattice. It corresponds to the vibration of LiC tetrahedra in the local environment model.

Information on the structural quality of the PLD LiMn$_2$O$_4$ films can be given considering th Raman data using the shape and the frequency of two groups of peaks located in the low- an high-frequency region of the spectra. The effect of target composition is clearly observed in Fig. When the PLD films are grown from target with Li$_2$O•10%, the RS spectra are less resolved an the peak at 622 cm^{-1} is broadened toward the high wavenumber side. This phenomenon is due t the amorphous structure of the films and the high distortion of MnO$_6$ octahedra. Analogous targe composition dependence has been observed on the Raman spectra of LiCoO$_2$ films grown PL [17].

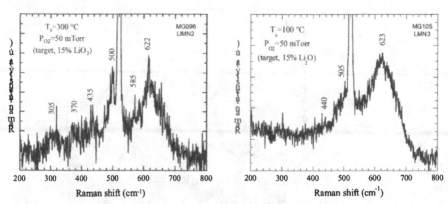

Fig. 3. Effect of the substrate temperature on the Raman spectra of PLD LiMn$_2$O$_4$ films.

The oxygen partial pressure in the deposition chamber influences the quality of the film. Th RS spectrum of the LIMN2 film grown at P(O$_2$)=50 mTorr corresponds to a highly disordere LiMn$_2$O$_4$ phase, while that of the LIMN5 film deposited at P(O$_2$)=100 mTorr exhibits well-resolve bands. The RS spectra of LiMn$_2$O$_4$ films deposited from the LiMn$_2$O$_4$+15% Li$_2$O target with a oxygen partial pressure of P(O₂)=50 mTorr as a function of substrate temperature show interesting differences between the film grown onto substrate at T_s=300°C and at T_s=100°C. The peaks situated in the low-wavenumber region, which are observed in the Raman spectrum of films grown on heated substrate, disappear when the film is grown on cool substrate. The RS spectrum of the film grown T_s=100°C displays the amorphous nature of the layer. It is obvious that better crystallinity of the films is obtained as the substrate temperature increases (Fig. 3). The band located at 640 cm⁻¹ can also be an indication of the presence of another manganese oxide such as Mn_3O_4 or Mn_2O_3.

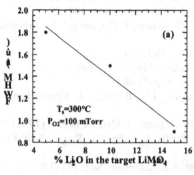 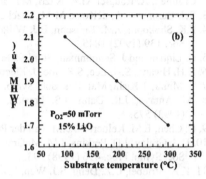

Fig. 4. Evolution of the FWHM of the symmetric stretching Raman mode of $LiMn_2O_4$ films as a function of (a) target composition and (b) substrate temperature (target $LiMn_2O_4$+15% Li_2O).

Fig. 4 shows the evolution of the full width at half maximum (FWHM) of the symmetric stretching Raman mode (622 cm⁻¹) of $LiMn_2O_4$ films as a function of (a) substrate temperature (target $LiMn_2O_4$+10% Li_2O), (b) substrate temperature (target $LiMn_2O_4$+15% Li_2O), and (c) target composition. It can be observed that low values of the FWHM for the Raman band located at 622 cm⁻¹ occurs for films grown at T_s=300°C in P(O₂)=100 mTorr from lithium-rich target. The RS spectrum of such a film is found to be well resolved, showing the polycrystalline nature of the films. As the stretching mode is sensitive to the film morphology, low FWHM values provide evidence for the spinel-like structure for samples grown at a high substrate temperature. These spectroscopic results indicate that the conjunction of target composition (lithium-rich), substrate temperature (T_s=300°C), and partial oxygen pressure (P(O₂)=100 mTorr) promotes reconstruction of the stoichiometric $LiMn_2O_4$ spinel framework.

CONCLUSION

X-ray diffraction data and RS measurements show that such a film crystallized in the spinel structure (Fd3m space group). Information for the structural quality of the PLD $LiMn_2O_4$ films can be given considering the Raman data using the shape and the frequency of two groups of peaks located in the low- and high-frequency region of the spectra. The effect of target composition is clearly observed. When the PLD films are grown from target with Li_2O•10%, the RS spectra are less resolved and the stretching Mn-O peak is broadened toward the high wavenumber side due to the high distortion of MnO_6 octahedra. These spectroscopic results indicate that the conjunction of target composition (lithium-rich), substrate temperature (T_s=300°C), and partial oxygen pressure (P(O₂)=100 mTorr) promotes reconstruction of the stoichiometric $LiMn_2O_4$ spinel framework.

ACKNOWLEDGMENTS

The authors thank Michel Lemal for his assistance in the x-ray diffraction measurements. Dr Saul Ziolkiewicz is greatfully acknowledged for his valuable comments.

REFERENCES

1. T. Ohzuku, in: *Lithium Batteries*, edited by G. Pistoia (Elsevier, Amsterdam, 1993), p. 239.
2. C. Julien, A. Rougier, G.A. Nazri, Mat. Res. Soc. Symp. Proc. 453 (1997) 647.
3. R.J. Gummow, A. deKock, M.M. Thackeray, Solid State Ionics 69 (1994) 59.
4. F.K. Shokoohi, J.M. Tarascon, B.J. Wilkens, D. Guyomard, C.C. Chang, J. Electrochem. Soc. 139 (1992) 1845
5. C. Liquan and J. Schoonman, Solid State Ionics 67 (1994) 17.
6. K.H. Hwang, S.H. Lee, S.K. Joo, J. Electrochem. Soc. 141 (1994) 3296.
7. T. Miura, T Kishi, Mat. Res. Soc. Symp. Proc. 393 (1995) 69.
8. M. Antaya, J.R. Dahn, J.S. Preston, E. Rossen, J.N. Reimers, J. Electrochem. Soc. 140 (1993) 575.
9. C. Chen, E.M. Kelder, P.J.J.M. van der Put, J. Schoonman, J. Mater. Chem. 6 (1996) 765.
10. D.S. Ginley, J.D. Perkins, J.M. McGraw, P.A. Parilla, M.L. Fu, C.T. Rogers, Mat. Res. Soc. Symp. Proc. 496 (1998) 293.
11. K.A. Striebel, C.Z. Deng, S.J. Wen, E.J. Cairns, J. Electrochem. Soc. 143 (1996) 1821.
12. A. Rougier, K.A. Striebel, S.J. Wen, E.J. Cairns, J. Electrochem. Soc. 145 (1998) 2975.
13. M. Morcrette, P. Barboux, J. Perriere, T. Brousse, Solid State Ionics 112 (1998) 249.
14. W.B. White, B.A. De Angelis, Spectrochim. Acta 23A (1967) 985.
15. C. Julien, C. Perez-Vicente, G.A. Nazri, Ionics 2 (1996) 1.
16. C. Julien, M. Massot, C. Perez-Vicente, E. Haro-Poniatowski, G.A. Nazri, A. Rougier, Mat. Res. Soc. Symp. Proc. 496 (1998) 415.
17. L. Escobar-Alarcon, E. Haro-Poniatowski, J. Jimenez-Jarquin, M. Massot and C. Julien, Mat. Res. Soc. Symp. Proc. 548 (1999) 223.

Mat. Res. Soc. Symp. Vol. 617 © 2000 Materials Research Society

SURFACE MORPHOLOGY OF PZT THIN FILMS PREPARED BY PULSED LASER DEPOSITION

Masaaki Yamazato[1], Masamitsu Nagano[1], Tomoaki Ikegami and Kenji Ebihara
[1]Department of Chemistry and Applied Chemistry, Saga University, 1 Honjo-machi, Saga, 840-8502, JAPAN
Department of Electrical and Computer Engineering, Kumamoto University, Kumamoto, 860-8555, JAPAN

ABSTRACT

Ferroelectric $PbZr_{0.52}Ti_{0.48}O_3$(PZT)/$YBa_2Cu_3O_{7-x}$(YBCO) heterostructures on MgO substrate were fabricated by KrF pulsed laser deposition. The grid electrode was set between a substrate and target for improvement of surface morphology. The typical PZT layer had excellent ferroelectric properties; remnant polarization of 39 $\mu C/cm^2$, coercive electric field of 41 kV/cm, loss tan δ =0.04, and dielectric constant of 950. X-ray diffraction results show that the films had highly c-axis and (a, b) plane orientation. The full widths at half-maximum (FWHM) of rocking curves was decreased with increasing the applied voltage of grid electrode. Atomic force microscopy (AFM) images of PZT layer showed that the film morphology was improved by using a grid electrode.

INTRODUCTION

Ferroelectric thin films such as $PbTiO_3$, $Pb(Zr, Ti)O_3$ (PZT) and $(Pb, La)(Zr, Ti)O_3$ are widely considered as high permittivity materials in dynamic random access memories (DRAMS), in ferroelectric random access memories (FERAMS) and the gate dielectric candidates in metal/ferroelectric/semiconductor field-effect transistors (MESFETS) [1,2]. Moreover, in recent years, electron emission from ferroelectric materials triggered by pulsed electric field has attracted much attention from the view point of various applications such as cold cathodes and emissive plasma display panels [3,4]. PZT/$YBa_2Cu_3O_{7-x}$(YBCO) heterostructures are interesting for various reasons. First, layered perovskite YBCO is compatible with perovskite PZT. Furthermore, it has been pointed out that degradation of remnant polarization due to cycling is significantly less when YBCO contact electrodes are used instead of Pt metal electrode in nonvolatile memory systems. For these electric device applications, it is necessary to obtain thin films with good uniformity and smooth surface morphology.

Ferroelectric thin films are fabricated by various techniques such as reactive sputtering, chemical vapor deposition, sol-gel spin-coat and pulsed laser deposition (PLD). PLD is very attractive process for oxide thin film preparation which guarantees the congruent transfer of highly nonthermal eroded target material to the growing film. High initial rate of heating and energetic plasma beam result in high kinetic and internal excitation energies of ablated species assist film growth and promote surface solid state reactions. However, the PLD method has some problems such as existence of particles or droplets [5,6,7,8].

In this paper, we report on a preparation of epitaxial $PbZr_{0.52}Ti_{0.48}O_3$/$YBa_2Cu_3O_{7-x}$ heterostructures on MgO substrates and improvement of surface morphology by setting a grid electrode between a substrate and a target.

EXPERIMENT

The PZT and YBCO films were prepared by KrF excimer laser ablation. A schematic diagram of the pulsed laser deposition system is shown in Fig. 1. Lambda Physik LPX305icc KrF excimer laser (λ=248 nm, pulse duration of 25 ns, max. output of 850 mJ) beam is directed at rotating stoichiometric ceramic targets of $PbZr_{0.52}Ti_{0.48}O_3$ and $YBa_2Cu_3O_{7-x}$. The resulting plume of ablated material is caught on MgO substrates placed in front of the target and heated by a resistance heater (Neocera Flat Plate Heater SH2). The distance from a PZT target and YBCO target to the substrate was 50 mm. The grid electrode (Cu, hole size is about 0.5 mm^2) was inserted between a substrate and a target. The distance from a substrate and grid electrode was 15 mm. The applied voltage of grid electrode V_g can be changed from -100 V to $+100$ V. The YBCO layer was grown at a substrate temperature of 710 °C in a 200 mTorr oxygen partial pressure ambient and at a laser fluence of 2 J/cm^2 with pulse repetition rate of 5 Hz. The PZT layer is deposited at a substrate temperature of 550 °C, 100 mTorr oxygen pressure and laser fluence of 2 J/cm^2. Finally the whole heterostructure was annealed in an oxygen pressure of 600 Torr at 400 °C for 60 min and then cooled down to room temperature. In Table I, these experimental conditions are listed. To measure the ferroelectric properties, circle (ϕ =200 µm) electrodes have been prepared *ex situ* by thermal evaporation of gold through the contact mask onto the front PZT/YBCO surface at room temperature. The deposited films were characterized using X-ray diffraction (XRD, Rigaku RINT2100/PC) and atomic force microscopy (AFM, SII SPI-3800N). Ferroelectric properties measured by using a Sawyer-Tower method and a LCZ meter (NF, 2345).

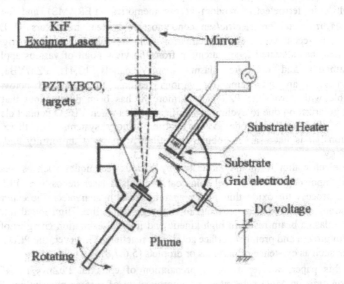

Figure 1. Schematic diagram of pulsed laser deposition system with grid electrode

Table 1. Preparation conditions

	YBCO	PZT
Laser fluence	2 J/cm^2	2 J/cm^2
Repetition rate	5 Hz	5 Hz
Laser shots	9,000	6,000-10,000
O$_2$ Pressure	200 mTorr	100 mTorr
Substrate	MgO (5 mm × 10 mm)	
Substrate temperature T_s	710 °C	550 °C
Applied voltage of grid electrode V_g	0 V	-100, 0 +100 V
Annealing	O$_2$ 600 Torr, 400 °C, 60 min	

RESULTS AND DISCUSSION

We have already reported on the structural, polarization and switching characteristics of PZT/YBCO/(MgO, Nd:YAlO$_3$) thin films by using a conventional PLD method (without grid electrode) [9,10]. In Ref. 5 and 6, the PZT/YBCO heterostructure showed high c-axis orientation, strong in (a, b) plane texture, high dielectric constant ε =950, low loss tan δ= 0.04, remnant polarization P_r = 32 µC/cm^2 and coercive electric field E_c = 41 kV/cm. Figure 2 shows the typical P-E hysteresis loop measured at room temperature (300 K) for the PZT film (prepared at grid voltage = 0V). The remnant polarization P_r = 39.0 µC/cm^2 and coercive electric field E_c = 41 kV/cm. In the case of using a grid electrode, P_r is higher than conventional PLD method. Ferroelectric property was improved by using a grid electrode. Figure 3 shows the applied voltage of grid electrode V_g dependence of P_r and E_c. The P_r is increased and E_c is decreased with increasing a V_g. At the V_g=100V, P_r and E_c are 41 µC/cm^2 and 39 kV/cm, respectively.

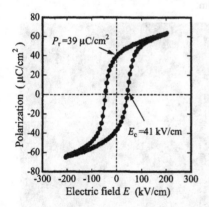

Figure 2. P-E hysteresis loop for a typical PZT(500nm)/YBCO(400nm)/MgO hetero-structure (prepared at V_g = 0V)

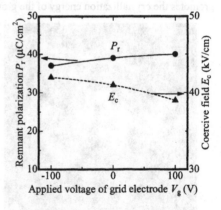

Figure 3. V_g dependence of remnant polarization P_r and coercive field E_c.

Figure 4(a) and 4(b) are the AFM images from the surface of the PZT films deposited by conventional PLD and by PLD with grid electrode ($V_g = +100$ V). Both films had the same film thickness of about 500 nm. As shown in Fig. 4, the roughness of the film grown by conventional PLD is about 40 nm, while that of the film grown by PLD with grid electrode is less than 20 nm. The density of particles and droplets ($\geq$ 500 nm) for PZT film grown by conventional PLD is $8 \times 10^6 / cm^2$, while that is reduced to $2 \times 10^6 / cm^2$. These results showed the surface morphology was improved by using a grid electrode. This is due to that the particles and droplets that travel to a substrate are prevented by grid electrode. However, the film growth rate is reduced to 15 nm/min, which is 60 % of that in conventional PLD method. The mechanism of this growth is that almost heavy particles and droplets in the plume are shielded by the grid electrode, while small particles can reach the substrate through the grid electrode. There is no dependence of grid electrode material. Kinoshita *et al* has reported that the same tendency by using the "eclipse method" PLD. In the "eclipse method", a shadow mask is positioned between a target and a substrate [6].

The high degree of *c*-axis orientation in both PZT and YBCO layers is observed by θ–2θ X-ray diffraction pattern and rocking curves. Furthermore, fabricated heterostructures are highly oriented in the (*a*, *b*) plane, so the perfect crystalline matching of PZT and YBCO layers is obtained. Figure 5 shows the rocking curves of PZT-002 reflection. The diffraction intensity is increased with increasing an applied voltage of grid electrode V_g. Their full widths at half-maximum (FWHM) and the calculated lattice parameter are showed in Figure6. The FWHM and *c*-axis lattice constant is decreased with increasing a V_g and the FWHM reaches a minimum value of 0.90°. In the case of PZT film grown by conventional PLD method, the FWHM is 1.15°. These results show that the crystalline property of PZT films has been improved by applying the positive voltage to grid electrode. Several authors reported that the active oxygen is superior by several orders of magnitude to the molecular oxygen, and that the molecular oxygen has little capability of forming good perovskite structures [11, 12, 13, 14]. One possible interpretation of improved crystallinity is that the active oxygen ions are accelerated by electric field, which promotes the crystallization energy of the growing film.

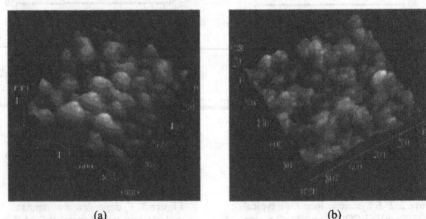

(a)   (b)

Figure 4. AFM images from the PZT films deposited by conventional PLD (a) and by PLD with grid electrode (b) ($V_g = +100$ V).

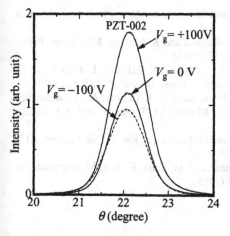

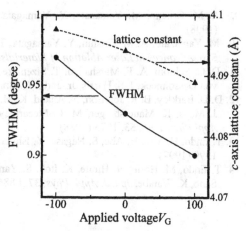

Figure 5. XRD rocking curves of PZT-002 reflection.

Figure.6 FWHM and c-axis lattice constant of PZT thin film.

CONCLUSIONS

The epitaxial $PbZr_{0.52}Ti_{0.48}O_3/YBa_2Cu_3O_{7-x}$ thin film heterostructures have been grown on the MgO single crystal substrate using the KrF PLD method. The grid electrode was set between a substrate and a target. The remnant polarization increased with increasing a bias voltage of grid electrode and reached a high value of 40 $\mu C/cm^2$. The ablated particles were selected according to size by grid electrode. The heavy particles and droplets in the plume were shielded by the grid electrode, while small particles could reach the substrate through the grid electrode. The density of droplets was reduced to $2\times10^6/cm^2$, which was 25 % of that in conventional PLD in the same O_2 ambient. The applied electric field of grid electrode improved the film crystallinity. The surface morphology of PZT grown by PLD with grid electrode was smoother than that grown by conventional PLD method.

REFERENCES

1. J. F. Scott and C. Araujo, *Science*, **246**, 1400 (1989).
2. D. J. Taylor, P. K. Larsen, and R. Cuppens, *Appl. Phys.Lett.*, **64**, 1392 (1994).
3. O. Auciello, M. A. Ray, D. Palmer, J. Duarte, G. E. McGuire, and D. Temple, *Appl. Phys. Lett.*, **66**, 2183 (1995).
4. Y. Kuratani, S. Omura, M. Okuyama and Y. Hamakawa, *Jpn. J. Appl., Phys.* **32**, 5471 (1995)
5. T. Arakawa, N. Arai, H. Yin, H. Kaneda, M. Sugahara and N. Haneji, Jpn. J. Appl. Phys., **38**, 2869 (1999)
6. K. Kinoshita, H. Ishibashi and T. Kobayashi, *Jpn. J. Appl. Phys.*, **33**, L417 (1994).
7. D-. W. Kim, S. –M. Oh, S. –H. Lee and T. W. Noh, *Jpn. J. Appl. Phys.*, **37**, 2016 (1998)
8. M. Tachiki, T. Hosomi and T. Kobayashi, *Jpn. J. Appl. Phys.*, **39**, 1817 (2000)

9.   A. M. Grishin, M. Yamazato, Y. Yamagata and K. Ebihara, *Appl. Phys. Lett.*, **72**, 620 (1998).

10. M. Yamazato, A. M. Grishin, Y. Yamagata, T. Ikegami and K. Ebihara, *MRS Symp. Proc.*, **526**, *Advances in Laser Ablation of Materials*, 187 (1998).

11. D. G. Schlom, A. F. Marshall, J. T. Sizemore, Z. J. Chen, J. N. Eckstein, I. Bozovic, K. E. von Dessonneck, J. S. Harris, Jr. and J. C. Bravman, *J. Cryst. Growth*, **102**, 361 (1990)

12. D. D. Barkley, B. R. Johnson, N. Anand, K. M. Beauchamp, L. E. Conroy, A. M. Goldman, J. Maps, K. Mauersberger, M. L. Mecartney, J. Morton, M. Tuominen and Y-J. Zhang, *Appl.Phys. Lett.*, **53**, 1973 (1988)

13. T. Endo, H. Yan, K. Abe, S. Nagase, Y. Ishida and H. Nishiku, *J. Vac. Sci. & Technol*, **A15**, 1990 (1997)

14. T. Endo, M. Horie, N. Hirate, K. Itoh, S. Yamada and M. Tada, K. Itoh, M. Sugiyama, S. Sano, K. Watabe, *Jpn. J. Appl. Phys*, **37**, L886 (1998)

Mat. Res. Soc. Symp. Vol. 617 © 2000 Materials Research Society

LOW TEMPERATURE GROWTH OF BARIUM STRONTIUM TITANATE FILMS BY ULTRAVIOLET-ASSISTED PULSED LASER DEPOSITION

V. CRACIUN, J. M. HOWARD, E. S. LAMBERS, R. K. SINGH
Department of Materials Science and Engineering, University of Florida, Gainesville, FL 32611

ABSTRACT

Barium strontium titanate (BST) thin films were grown directly on Si substrates by an *in situ* ultraviolet (UV)-assisted pulsed laser deposition (UVPLD) technique. With respect to films grown by conventional (i.e. without UV illumination) pulsed laser deposition (PLD), the UVPLD grown films exhibited improved structural and electrical properties. The dielectric constant of a 40-nm thick film deposited at 650 °C was determined to be 281, the leakage current density was approximately 4×10^{-8} A/cm^2 at 100 kV/cm, and the density of interface states at the flat-band voltage was found to be approximately 5.6×10^{11} eV^{-1}cm^{-2}. X-ray photoelectron spectroscopy investigations found that the surface of the grown films exhibited an additional Ba-containing phase, besides the usual BST perovskite phase, which was likely caused by stress and/or oxygen vacancies. The amount of this new phase was always smaller and very superficial for UVPLD grown films, which can explain their better overall properties.

INTRODUCTION

Barium strontium titanate, $Ba_xSr_{1-x}TiO_3$ (BST), possesses a high dielectric constant ($\varepsilon_r \gg \varepsilon_{SiO_2}$) [1-4], which makes it attractive for capacitor applications. Several studies have also suggested that BST could be a potential replacement of SiO_2 thin layers in future MOS circuits [5-7]. The results obtained so far have shown that thin BST layers usually exhibit much lower dielectric constants than the bulk value [8]. Fine grain microstructure, high stress levels, the presence of oxygen vacancies, the formation of interfacial layers, and the oxidation of the bottom electrode or Si are believed to be factors which can contribute to the degradation of the electrical properties [9-11]. In addition to the interfacial layer, which is formed due to the high temperatures (700-800 °C) and oxygen pressures (~200 mTorr) required for growth of high quality BST films, it has been found that Ba atoms on the surface of the film are present in two chemical states [12]. Besides the usual BST perovskite phase, an additional phase was found, most likely caused by the presence of mechanical stresses and/or oxygen vacancies. This phase will also negatively affect the electrical characteristics of the grown structure.

One of the most promising methods for the growth of high quality thin BST films, which could alleviate some of the problems mentioned above, is pulsed laser deposition (PLD) [13-17]. In this paper we demonstrate the growth of BST films directly on Si wafers of very low equivalent SiO$_2$ thickness (<10 Å) employing an *in situ* ultraviolet (UV)-assisted pulsed laser deposition (UVPLD) technique. The results are compared with those obtained from conventional PLD grown films.

EXPERIMENT

$Ba_{0.5}Sr_{0.5}TiO_3$ targets were ablated using a KrF excimer laser system in a typical PLD setup which was described in detail elsewhere [18, 19]. The deposition parameters used were 450 mJ laser energy (fluence ~ 2 J/cm^2), 5 Hz repetition rate, and 250 mTorr and 150 mTorr oxygen pressure for conventional PLD and UVPLD, respectively. The films were deposited at 450, 550, and 650 °C for different times on Si substrates that were rinsed in acetone, methanol, and 1% HF, blown dry by N_2 and then immediately loaded into the deposition chamber. To enhance oxygen incorporation into the growing films during the deposition, a vacuum compatible, low pressure Hg lamp was fitted into the PLD system. It has a fused silica envelope, which allows more than 85% of the emitted 184.9 nm radiation (around 6% of the output), to be transmitted.

The crystallinity of the films was examined by x-ray diffraction (XRD). The chemical composition and bonding were investigated by x-ray photoelectron spectroscopy (XPS) in a Perkin Elmer 5100 installation using Mg Kα radiation (1253.6 eV). Fourier transform infrared spectroscopy (FTIR) was used to identify the presence of any carbonate layers on samples. The thickness and optical properties of the films were measured by spectroscopic ellipsometry at 70°. The capacitance-voltage characteristics for MOS capacitors were measured with a Hg probe at 100 kHz using a HP4275A LCR meter. The current-voltage (I-V) characteristics were determined by depositing Au contacts onto BST films and measuring the leakage current of the formed structures with a semiconductor parameter analyzer.

RESULTS AND DISCUSSION

The XRD patterns acquired from BST layers grown by UVPLD and conventional PLD techniques at various temperatures showed the presence of randomly oriented crystalline grains (perovskite phase [14-16]). The UVPLD films always showed smaller full width at half maximum (FWHM) diffraction peaks, pointing to a larger grain size than that found in the PLD film. This can be explained by the increase of the surface mobility of adatoms by UV irradiation and the presence of more reactive atomic oxygen and ozone species [20, 21].

Results of the XPS investigations of the surface of as-grown BST layers by UVPLD and PLD are presented in figures 1a and 1b, respectively. Each of the Ba $3d_{3/2}$ and Ba $3d_{5/2}$ peaks can be fitted by two peaks, separated by approximately 1.4 eV. The lower binding energy peaks (denoted Ba1), located at 794.2 eV and 779.0 eV, correspond to Ba atoms bound into a BST perovskite phase. The peaks located at 797.3 eV and 781.2 eV (denoted Ba2) were initially believed to correspond to Ba atoms within a decomposed surface barium carbonate layer [12, 22]. However, FTIR investigations did not show the presence of CO_3^{2-} absorption bands [23]. Even for films grown at 450 °C, where most of the surface Ba atoms were found to belong to this new phase, there was no evidence of a carbonate absorption band at 1450 cm^{-1}. It is thus suggested that this new phase should consist of a stressed BST layer, probably caused by strain and/or oxygen vacancies [24]. The presence of this new phase can cause a decrease in the measured dielectric constant value of the BST layer, especially for very thin films, where its relative importance is greater.

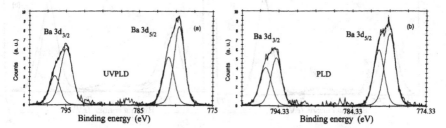

Figure 1. Acquired and fitted XPS spectra of the Ba 3d region for (a) UVPLD and (b) PLD as-grown BST films; take-off angle was 45°.

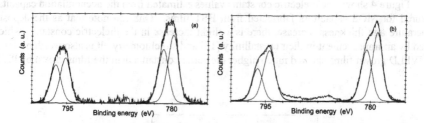

Figure 2. Acquired and fitted XPS spectra of the Ba 3d region for (a) UVPLD and (b) PLD as-grown BST films; take-off angle was 20°.

Comparing the XPS spectra displayed in Fig. 1, one can note that the UVPLD grown sample exhibited a smaller fraction (35% versus 45%) of Ba atoms within this new phase. Spectra acquired at a take-off angle of 20° to enhance the surface contribution are displayed in figure 2. One can see that while for the PLD grown sample the ratio of Ba1 to Ba2 peak intensities remained almost constant, at around ~41%, it increased significantly for the UVPLD grown sample, from ~30% to ~44%. This suggests that the surface layer of PLD grown film consists of a rather homogeneous mixture of these two phases which extends more than 4.0 nm inside the sample depth, while it is only a very thin film (~1 nm) uniformly covering the BST phase for the UVPLD grown sample. Also, analyzing the O 1s photoelectron peak acquired from the same samples it was found that the amount of oxygen atoms bound to BST phase was several percent higher in the UVPLD grown sample than in the PLD one. This seems to support the idea that this new surface-located phase is caused by oxygen deficiencies.

XPS spectra acquired from samples that were sputter-cleaned with Ar ions (6 min, 4 keV), until no C contamination was found on the surface, showed a subtle difference between the shape of the Ba 3d peaks. For PLD grown samples, a small Ba suboxide peak located at lower binding energy, as seen in figure 3, appeared. As a 4 keV Ar ion sputtering could not break Ba-O bonds of a stoichiometric BST film [25], it is therefore suggested that the outermost layer should already contain some defects and oxygen vacancies. It thus appears that the more reactive oxygen atmosphere present during the UVPLD process results in a more stoichiometric and stable BST layer.

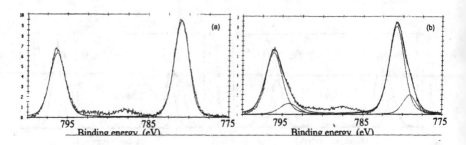

Fig. 3. Acquired and fitted XPS spectra of the Ba 3d region for (a) UVPLD and (b) PLD grown BST films after 4 keV Ar ion sputtering for 6 min; take-off angle was 20°.

Figure 4 shows the dielectric constants values estimated from the accumulation capacitance measured for MOS capacitors fabricated from BST films. One can note that as the deposition temperature and thickness increase, there is a slight increase in the dielectric constants, which is caused by an improvement in their crystallinity [15] and stoichiometry. It is also worth noting that the UVPLD grown films showed much higher dielectric constants than the films grown by PLD.

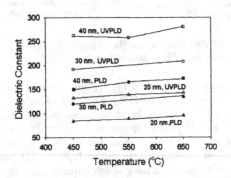

Figure 4. Dielectric constant values of BST films grown by PLD or UVPLD at different substrate temperatures.

The electrical parameters such as flat band voltage (V_{fb}), density of states (D_{it}), and equivalent oxide thickness (T_{eq}) extracted from the C-V characteristics of the films are summarized in Table I. The beneficial UV effect upon the electrical properties of the grown films could be seen even for the smallest thickness (~20 nm) and lowest deposition temperatures (~450 °C) employed during this study. The equivalent oxide thickness values obtained for the best UVPLD films were around 10 Å, which is comparable to the values obtained recently by McKee et al. [26] where the films were grown by molecular beam epitaxy. Similarly the interface trap density data listed in Table I show that UVPLD grown BST films exhibited much lower values.

The interface trap density for a 20-nm thick BST deposited by UVPLD was around 5.6×10^{11} eV^{-1}cm^{-2}, which is also comparable to best values reported in the literature [27]. The values listed in Table I suggest that UV irradiation during the ablation growth process significantly helped to improve MOS capacitor characteristics.

TABLE I. The electrical parameters of the BST films

Temp. (°C)	Thickness (nm)	T_{eq} (nm)		V_{fb} (V)		D_{it} (eV^{-1}cm^{-2})	
		UVPLD	PLD	UVPLD	PLD	UVPLD	PLD
450	20	1.06	1.68	+1.01	+1.12	7.1×10^{11}	2.1×10^{12}
450	30	1.10	1.75	+1.02	+1.09	7.8×10^{11}	2.9×10^{12}
450	40	1.06	1.86	+1.05	+1.10	8.1×10^{11}	4.3×10^{12}
550	20	1.00	1.57	+1.02	+1.15	6.6×10^{11}	2.0×10^{12}
550	40	1.08	1.69	+1.02	+1.15	7.9×10^{11}	3.4×10^{12}
650	20	0.98	1.47	+0.80	+1.15	5.6×10^{11}	1.8×10^{12}
650	30	1.01	1.56	+1.00	+1.10	6.9×10^{11}	2.2×10^{12}
650	40	0.99	1.62	+1.02	+1.12	7.2×10^{11}	2.3×10^{12}

At a field of 100 kV/cm 40 nm thick UVPLD-grown films deposited at 650 °C by UVPLD exhibited a leakage current of 6×10^{-8} A/cm^2, which was nearly a factor of 1.5 better than that measured for PLD deposited films. Such values are among the best reported so far for BST films [28]. Although the leakage current mechanism involved in this case is not clear, we expect it to be a Schottky conduction mechanism [29], as the current is almost constant until 750 kV/cm beyond which it shows a steep increase.

CONCLUSIONS

In summary, we have shown that *in situ* UV irradiation, which exposes each grown atomic layer to the action of energetic photons and reactive oxygen gaseous species during laser ablation-deposition, resulted in the growth of high quality BST at moderate substrate temperatures. These films exhibited better electrical properties than conventional PLD grown films under similar conditions. XPS investigations showed that UVPLD grown films contained less Ba atoms inside a stressed layer that was formed on the surface. By using this *in situ* UVPLD technique, MOS capacitors with an equivalent oxide thickness of less than 10 Å and very low leakage currents have been obtained at moderate deposition temperatures

ACKNOWLEDGMENTS

We thank A. Srivastava for his help with electrical characterization.

REFERENCES

1. H. Komiya, IEEE Symposium on VLSI Technology, 1 (1997).
2. G. Arlt, D. Hennings, and G. de With, J. Appl. Phys. **58**, 1619 (1985).

3. T. M. Shaw, Z. Suo, M. Huang, E. Liniger, R. B. Laibowitz, and, and J. D. Baniecki, Appl. Phys. Lett. **75**, 2129 (1999).

4. L. Kinder, X. F. Zhang, I. L. Grigorov, C. Kwon, Q. X. Jia, L. Luo, and J. Zho, J. Vac. Sci. Technol. A **17**, 2148 (1999).

5. L. C. Feldman, E. P. Gusev, and E. Garfunkel in Fundamental Aspects of Ultrathin Dielectrics on Si-based Devices, edited by E. Garfunkel et al. (Kluwer, Dordrecht, 1998).

6. J. D. Meindl, J. Vac. Sci. Technol. **14**, 192 (1996).

7. National Technology Roadmap for Semiconductors (Sematech Corp., San Jose, CA, 1997).

8. W. Chang, J. S. Horwitz, A. C. Carter, J. M. Pond, S. W. Kirchoefer, C. M. Gilmore, and D. B. Chrisey, Appl. Phys. Lett. **74**, 1033 (1999).

9. T. M. Shaw, Z. Suo, M. Huang, E. Liniger, R. B. Laibowitz, and J. D. Baniecki, Appl. Phys. Lett. **75**, 2129 (1999).

10. Y. Gao, A. H. Mueller, E. A. Irene, O. Auciello, A. Krauss, and J. A. Schultz, J. Vac. Sci. Technol. A **17**, 1880 (1999).

11. H.-F. Cheng, J. Appl. Phys. **79**, 7965 (1996).

12. Y. Fujisaki, Yasuhiro Shimamoto, and Yuichi Matsui, Jpn. J. Appl. Phys. **38** pt. 2, L52 (1999).

13. F. Tcheliebou and S. Baik, J. Appl. Phys. **80**, 7046 (1996).

14. R. Thielsch, K. Kaemmer, B. Holzapfel, and L. Schultz, Thin Solid Films **301**, 203 (1997).

15. H.-F. Cheng, J. Appl. Phys. **79**, 7965 (1996).

16. S. Saha and S. B. Krupanidhi, Mat. Sci. Eng. **B57**, 135 (1999).

17. L. A. Knauss, J. M. Pond, J. S. Horwitz, D. B. Chrisey, C. H. Mueller, and R. Treece, Appl. Phys. Lett. **69**, 25 (1996).

18. A. Srivastava, V. Craciun, J. Howard, and R. K. Singh, Appl. Phys. Lett. **75**, 3002 (1999).

19. A. Srivastava, D. Kumar, and R. K. Singh, Electrochem. Solid State Lett. **2**, 294 (1999).

20. V. Craciun, J. Howard, and R. K. Singh, presented at MRS Spring Meeting, San Francisco, 6-10 April 1999.

21. V. Craciun and R. K. Singh, Electrochem. Solid-State. Lett. **2**, 446 (1999).

22. C. Miot, E. Husson, C. Proust, R. Erre, and J. P. Coutures, J. Mater. Res. **12**, 2388 (1997).

23. M. C. B. Lopez, G. Fourlaris, B. Rand, and F. L. Riley, J. Am. Ceram. Soc. **82**, 1777 (1999).

24. S. M. Mukhopadhyay and T. C. S. Chen, J. Mater. Res. **10**, 1502 (1995).

25. D. Leinen, A. Fernandez, J. P. Espinos, A. R. Gonzalez-Elipe, Appl. Phys. A**63**, 237 (1996).

26. R. A. McKee, F. J. Walker, and M. F. Chisholm, Phys. Rev. Lett. **81**, 3014 (1998).

27. M. Alexe, Appl. Phys. Lett. **72**, 2283 (1998).

28. J. M. Lee, S. Y. Kang, J. C. Shin, W. J. Kim, C. S. Hwang, and H. J. Kim, Appl. Phys. Lett. **74**, 3489 (1999).

29. C. S. Hwang, B. T. Lee, C. S. Kang, J. W. Kim, k. H. Lee, H. J. Cho, H. Horii, W. D. Kim, S. I. Lee, Y. B. Roh, and M. Y. Lee, J. Appl. Phys. **83**, 3703 (1998).

Mat. Res. Soc. Symp. Vol. 617 © 2000 Materials Research Society

TANTALUM NITRIDE THIN FILMS SYNTHESIZED BY PULSED Nd:YAG LASER DEPOSITION METHOD

Hiroharu Kawasaki, Kazuya Doi, Jun Namba and Yoshiaki Suda

Department of Electrical Engineering, Sasebo National College of Technology, Okishin 1-1, Sasebo, Nagasaki, 857-1193, Japan

ABSTRACT

Tantalum nitride (TaN) films have been deposited on silicon substrates by using a pulsed Nd:YAG laser deposition method. Experimental results suggest that the substrate temperature is one of the most important parameters to prepare crystalline tantalum nitride thin films. Glancing-angle X-ray diffraction patterns show that the films deposited at Ts $\leq$ 300 °C are almost amorphous, and crystalline $Ta_6N_{2.57}$ films are obtained at Ts $\geq$ 500 °C. Grain size of the film increases with increasing substrate temperature.

INTRODUCTION

Tantalum nitride (TaN) has been extensively used in many applications due to some interesting properties, such as high hardness and high temperature stability[1,2]. In the silicon technology, copper (Cu) has drawn much attention recently as a new interconnecting material for deep submicron integrated circuits as a replacement for aluminium (Al) and its alloys. However, Cu is known to diffuse into either Si or SiO_2 layers during processing, and this problem is addressed by employing a suitable diffusion barrier layer between the copper overlayer and the silicon substrate. In this direction, TaN and Ta_2N compounds have been investigated and found to be promising materials to serve as a diffusion barrier because TaN-based materials can withstand high temperatures[3-6]. As these properties, TaN films have been deposited by several physical and chemical vapor deposition techniques. However, high quality TaN films for the diffusion barrier layer have not been prepared.

Pulsed Nd:YAG laser deposition (PLD) method is a widely used technique for the deposition of thin films, due to the advantages of a simple system setup, wide ranging deposition conditions, wider range choice of materials and higher instantaneous deposition rate. Because of this versatility, we have developed several kinds of functional thin films, such as tungsten carbide[7,8], silicon carbide[9], chromium carbide[10], cubic boron nitride[11], carbon nitride[12] and silicon nitride[13] using the PLD method. In this paper, we describe the fundamental characteristics of the tantalum nitride (TaNx) films prepared using the PLD method as well as the conditions required to fabricate crystalline and amorphous tantalum nitride thin films.

EXPERIMENTAL

The schematic of the experimental apparatus is shown in Fig. 1. A deposition chamber was fabricated of stainless steel with a diameter of 400 mm and a length of 370 mm. The chamber was evacuated to a base pressure (below 4.0×10^{-4} Pa) using a turbo molecular pump and a rotary pump. A pulsed Nd:YAG laser (Lumonics YM600; wavelength of 532 nm, pulse duration of 6.5 ns, maximum output energy of 340 mJ) was used to irradiate $Ta_6N_{2.57}$ (purity $\geq$ 99 %) targets. Their radiated area was maintained at 2.8 mm^2. The laser energy density (Ed) was fixed at 3.8 J/cm^2. The targets were rotated at 20 rpm to avoid pitting during deposition.

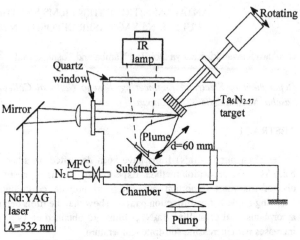

Fig. 1 Schematic diagram of the experimental apparatus.

The Si(100) substrates were cleaned using an ultrasonic agitator by repeated bathing in ethanol and then rinsed in high-purity deionized water prior to loading into the deposition chamber. The substrates were located at a distance of 60 mm from the facing target and were heated to ~750°C using an IR lamp. After 18000 laser pulses at a 10 Hz repetition rate, the deposition process was completed. Details of the preparation conditions of TaNx thin films are given in Table I.

The surface morphology was observed using a field-emission scanning electron microscope (FE-SEM; JEOL JSM-6300F). The film thickness, measured by α-step (KLA Tencor; AS500), was about 75 nm and the growth rate was 2.5 nm/min. The crystalline structure and crystallographic orientation of the films were characterized by glancing-angle X-ray diffraction (GXRD; PHILIPS PW1350)) using Cu Kα radiation where the angle of incidence was kept at 1.0°.

RESULTS AND DISCUSSION

Surface morphology of the TaNx thin films on Si(100) substrates was examined using

Table I   Preparation conditions

Laser	Pulsed Nd:YAG laser
	Wavelength 532 nm
	Pulsed width 6.5 ns
	Energy density 3.8 J/cm^2
	Repetition rate 10 Hz
Target	$Ta_6N_{2.57}$ (purity 99%)
Rotating speed	20 rpm
Substrate	Si(100)
Substrate temp.	room temp ~ 750 °C
Distance	d = 60 mm
Base pressure	$< 4.0 \times 10^{-4}$ Pa
Gas pressure	Base pressure
Deposition time	30 min

FE-SEM at various substrate temperatures (Ts). Figure 2(a) shows the micrograph of the film prepared at Ts = room temperature (R.T.). A lot of 50~200 nm-size clusters are superimposed on the surface of the films. Fig. 2(b) shows the micrograph of the film grown at Ts = 750°C. The sizes of these clusters increase with increasing Ts and become about 500~800 nm.

Detailed GXRD measurements were conducted to study the crystalline properties of the deposited TaNx films. Figure 3 shows the GXRD pattern of a reference target pellet of $Ta_6N_{2.57}$ in which numerous crystalline peaks of $Ta_6N_{2.57}$ were identified. The film deposited at Ts = R.T.

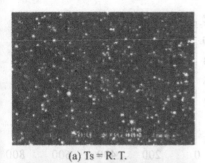

(a) Ts = R. T.

(b) Ts = 750 °C

Fig. 2 FE-SEM micrographs of the TaNx films deposited at (a) Ts = R. T. and (b) Ts = 750 °C ($P = 4 \times 10^{-4}$ Pa, Ed = 3.8 J/cm^2, d = 60 mm).

is amorphous, because no peak is observed in Fig. 4(a). When the substrate temperature is increased to 200~500°C, several indistinct peaks appear around the scattering angle of 35~45°, as shown in Figs. 4(b) and 4(c). When the temperature is increased to 700°C, distinct peaks of crystalline $Ta_6N_{2.57}$(110) (002) (111) and (300) appear, as shown in Fig. 4(d). This GXRD pattern is consistent with that of the $Ta_6N_{2.57}$ target shown in Fig. 3. To estimate the crystallinity grade of the prepared films, the crystalline grain size was determined from the Full Width Half Maximum (FWHM) of the X-ray peaks, as shown in Fig. 4.

$$\text{grain size} = \frac{0.9 \cdot \lambda}{\cos\theta \cdot \text{FWHM}} \qquad (1)$$

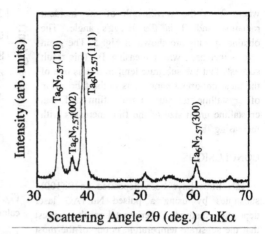

Fig. 3 GXRD pattern of CuKα obtained from a $Ta_6N_{2.57}$ target.

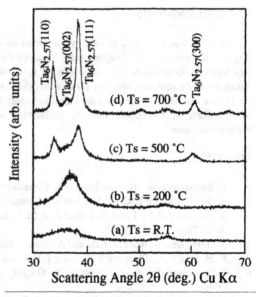

Fig. 4 GXRD patterns of TaNx films deposited at (a) Ts = R. T. , (b) Ts = 200 °C, (c) Ts = 500 °C and (d) Ts = 700 °C. ($P = 4 \times 10^{-4}$ Pa, Ed = 3.8 J/cm^2, d = 60 mm).

where λ is the wavelength of the incident radiation and θ is the Bragg's angle. The obtained results are shown in Fig. 5. The grain size is increased with increasing Ts. This result suggests that the substrate temperature is one of the most important parameters in the fabrication of crystalline tantalum nitride film and that crystalline grain size of the film increases with increasing Ts.

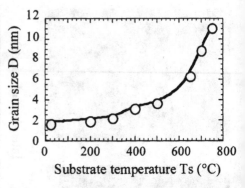

Fig. 5 Crystallinity grade of the prepared films calculated from equation (1).

CONCLUSIONS

TaN films have been deposited on Si(100) substrates by using a pulsed Nd:YAG laser deposition method. Experimental results suggest that the substrate temperature is one of the most important parameters to prepare crystalline tantalum nitride thin films. GXRD patterns show that all deposited TaNx films are almost amorphous at Ts $\leq$ 300 °C, and crystalline Ta$_6$N$_{2.57}$ films are obtained at Ts $\geq$ 500 °C. Crystalline grain size of the film increases with increasing the Ts.

ACKNOWLEDGMENTS

This work was supported in part by the Grant-in-Aid for Scientific Research (B) and the Regional Science Promoter Program and a Research Fund from the Nagasaki Super Technology Development Association. The authors wish to thank Drs. K. Ebihara, T. Ikegami and Y. Yamagata of Kumamoto University for their helpful discussions. The authors also wish to thank Dr. H. Abe and Mr. H. Yoshida of the Ceramic Research Center of Nagasaki, Drs. S. Aoqui and M. Munakata of Kumamoto Institute Technology for their technical assistance with the experimental data.

REFERENCES

1   K. Radhakrishnan, Ng Geok Ing and R. Gopalakrishnan: Mat. Sci. Eng. B57 224 (1999).
2   Y. Qin, L. Liu and L. Chen: J. Alloys and Compounds 269 238 (1998).
3   K, Holloway, P. M. Fryer. C. Cabral Jr, J. M. E. Harper, P. J. Bailey and K. H. Keileher: J. Appl. Phys. 71 5433 (1992).
4   M. Takeyama, A. Noya, T. Sasc and A. Ohta: J. Vac. Sci. Tcchnol. B 14 674 (1996).
5   K. H. Min, K. C. Chun and K. B. Kirn: J. Vac. Sci. Technol. B 14 3263 (1996).
6   J. O. Olowolafce, C. J. Mogab, R. B. Gregory and M. Kottke: J. Appl. Phys. 72 4099 (1992).
7   Y. Suda, T. Nakazono, K. Ebihara and K. Baba: Nucl. Instr. and Meth. in Phys. Res. B. 121 396 (1997).
8   Y. Suda, T. Nakazono, K. Ebihara, K. Baba and H. Hatada: Materials Chemistry and Physics 54 177 (1998).
9   Y. Suda, H. Kawasaki, R. Terajima, M. Emura, K. Baba, H. Abe, H. Yoshida, K. Ebihara

and S. Aoqui: J. Korea. Phys. Soc. **35** S88 (1999).

10 Y. Suda, H. Kawasaki, R. Terajima and M. Emura: Jpn. J. Appl. Phys. **38** 3619 (1999).
11 Y. Suda, T. Nakazono, K. Ebihara and K. Baba: Thin Solid Films **281-282** 324 (1996).
12 Y. Suda, T. Nakazono, K. Ebihara, K. Baba and S. Aoqui: Carbon **36** 771 (1998).
13 Y. Suda, K. Ebihara, K. Baba, H. Abe and A. M. Grishin: Nano Structured Materials **12** 291 (1999).

Mat. Res. Soc. Symp. Vol. 617 © 2000 Materials Research Society

Properties of the magnetoresistive La$_{0.8}$Sr$_{0.2}$MnO$_3$ film and integration with PbZr$_{0.52}$Ti$_{0.48}$O$_3$ ferroelectrics

Fumiaki Mitsugi[1], Tomoaki Ikegami[1], Kenji Ebihara[1], J. Narayan[2] and A. M. Grishin[3]

[1]Department of the Electrical and Computer Engineering, Kumamoto University, Kurokami, Kumamoto 860-8555, Japan

[2]Department of the Materials Science and Engineering, North Carolina State University, Raleigh, North Carolina 27695-7916, USA

[3]Department of Condensed Matter Physics, Royal Institute of Technology, S-10044 Stockholm, Sweden

ABSTRACT

The colossal magnetoresistive La$_{0.8}$Sr$_{0.2}$MnO$_3$ (LSMO) thin film was prepared on the MgO (100) single crystal substrate using KrF excimer pulsed laser deposition technique. The LSMO film deposited at the substrate temperature of 850 °C, oxygen pressure of 500 mTorr and laser energy density of 2 J/cm^2 (5 Hz) showed the resistivity peak temperature (T_p) of 330 K and the magnetoresistance change of 15 % (H=0.7 T) at the room temperature. The large lattice mismatch with the substrate increased T_p and decreased the resistivity of the LSMO film.

The X-ray diffraction measurement for the PbZr$_{0.52}$Ti$_{0.48}$O$_3$ (PZT) / LSMO heterostructures indicated both c-axis and in- plane orientation, with the good PZT surface morphology.

INTRODUCTION

Perovskite manganites of the Re$_{1-x}$Ae$_x$MnO$_3$ (Re : rare earth elements, Ae : alkaline earth elements) have potential for various device applications such as magnetic field sensor, hard disk read head and infrared bolometer due to the negative colossal magnetoresistance (CMR) effect. The basic manganite of ReMnO$_3$ is an antiferromagnetic insulator. A substitution of the divalent Ae^{2+} for the trivalent Re^{3+} makes Mn^{3+}(t^3_{2g} e$_g^1$) / Mn^{4+}(t^3_{2g}e$_g^0$) mixed valence state. The electron hopping from Mn^{3+} to Mn^{4+} by a Mn-O-Mn path results in ferromagnetic ordering and metallic conductivity. The electron (hole) transfer between the neighboring sites can be expressed as $t_{ij} = t_0 \cos (\theta_{ij} / 2)$ [1]. In this equation, θ_{ij} is the relative angle between the neighboring spins. The Mn^{3+} / Mn^{4+} ratio, Mn-O-Mn bond length and bond angle have an effect on the transfer property. The sizes of the Re^{3+} ion and Ae^{2+} ion change the Mn-O-Mn bond length and the many materials such as La$_{1-x}$Ca$_x$MnO$_3$, La$_{1-x}$Ba$_x$MnO$_3$, La$_{1-x}$Sr$_x$MnO$_3$, Nd$_{1-x}$Sr$_x$MnO$_3$ have been researched. We have studied La$_{1-x}$Sr$_x$MnO$_3$ system. The large radii of La^{3+} ion (1.15Å) and Sr^{2+} (1.13Å) ion shorten the Mn-O-Mn bond length, which results in high resistivity peak temperature (T_p) [2]. The bulk La$_{1-x}$Sr$_x$MnO$_3$ has ferromagnetic insulator – ferromagnetic metal transition at x=0.17 [3]. We report the properties of x=0.20 La$_{1-x}$Sr$_x$MnO$_3$ thin film. We have also tried to integrate the CMR film with ferroelectric film. The ferroelectrics / CMR heterostructures have a potential for not only ferroelectric field effect transistor, but also new device which works under both electric and magnetic field.

In this paper, we investigated the structural and electrical properties of the $La_{0.8}Sr_{0.2}MnO_3$ (LSMO) film deposited on MgO (100) substrate by KrF excimer laser deposition (PLD) method. The $PbZr_{0.52}Ti_{0.48}O_3$ (PZT) / LSMO heterostructures were also studied for understanding fundamental of the ferroelectric memory device.

EXPERIMENTAL DETAILS

The LSMO thin films were prepared from a stoichiometric target of $La_{0.8}Sr_{0.2}MnO_3$ by PLD technique on MgO (100) substrates. The ambient oxygen gas was fed into the stainless steel chamber (Φ=280 mm) after evacuating to base pressure of 10^{-6} Torr. The pressure of ambient gas was adjusted by a mass flow controller. The MgO (100) substrate was heated by a flat plate heater (Neocera). KrF excimer laser (Lambda Physik LPX305icc, λ=248 nm, pulse duration=25 ns, maximum output=850 mJ/pulse) was used and typical energy density and repetition rate were 2 J/cm^2 and 5 Hz, respectively. The deposition time was 10 min (3000 pulses) and the distance from the target to MgO substrate was 40 mm. The substrate temperature (T_s) and oxygen pressure (P_{O2}) were varied in the range of 650-850 °C and 100-500 mTorr, respectively.

The PZT film was prepared on the LSMO / MgO with a laser energy density of 2 J/cm^2 (5 Hz). The deposition time was 12.5 min (3750 pulses). T_s and P_{O2} were 550 °C and 100 mTorr, respectively. After the deposition of PZT, the whole structures (PZT / LSMO / MgO) were post annealed in 600 Torr oxygen at 400 °C for 60 min.

The crystalline structure was examined using X-ray diffractometer (XRD : Rigaku RINT2000/PC) with CuK$_\alpha$ radiation. The surface morphology of the film was observed by the atomic force microscopy (AFM : Seiko Instruments Inc. SPI3800N). The film thickness was estimated by the cross-sectional scanning electron microscopy (SEM : JEOL JSM-T200). The conventional four terminal resistance method was used to measure the resistivity versus temperature (ρ - T) curves and the magnetoresistive effect. The ferroelectric properties such as polarization (P) - electric field (E) hysteresis loop were measured by Sawyer-Tower circuit.

RESULTS AND DISCUSSION

LSMO thin film

We prepared the LSMO films on MgO substrate changing the deposition conditions of T_s (650-850 °C) and P_{O2} (100-500 mTorr). It is required that the T_p of CMR material is higher than room temperature to realize CMR complex devices. The higher values of T_s and P_{O2} increased T_p, decreased the electric resistivity and shortened c-axis lattice constant [4]. The sample deposited at T_s of 850 °C in P_{O2} of 500 mTorr showed the highest T_p of 330 K and shortest c-axis lattice constant of 3.88 Å. This result suggests that the short Mn-O-Mn bond length enhances the electron transfer energy. However, the short lattice constant results in the large lattice mismatch against the MgO substrate. Figure 1 shows the relationships between the lattice mismatch against MgO substrate and T_p, temperature coefficient of resistivity (TCR) peak temperature $T_{p(TCR)}$ and the maximum resistivity ρ_m of the LSMO film. The lattice constant of the MgO substrate is 4.21 Å. The LSMO has been considered to be pseudocubic crystal. It is shown that the large lattice mismatch increases T_p and decreases the resistivity. The fact suggests the strain generated in the film has a significant effect on the electron transfer. Khartsev et al. reported that the ultra thin

LSMO film under the strain due to the lattice mismatch increases T_p and decreases resistivity [5]. Figure 2 indicates the full width at the half maximum (FWHM) of the LSMO (002) rocking curve versus lattice mismatch. Increasing the lattice mismatch provides narrow rocking curve. Figure 3 shows the temperature dependence of the resistivity, TCR and magnetoresistance (MR) of the sample deposited at T_s of 850 °C in P_{O2} of 500 mTorr. The TCR and MR are defined as $(1/\rho)(d\rho / dT)$ and $-[\rho(H)-\rho(0)]/\rho(0)$, respectively. The MR peak temperature $T_{p(MR)}$ of 290 K appears at almost the same temperature as $T_{p(TCR)}$. The values of MR and TCR are 15% ($H=0.7$ T) and 3 %/K at 290 K. Table I is a summary of recent published results of colossal magnetoresistive $La_{1-x}Sr_xMnO_3$, $La_{1-x}Ca_xMnO_3$ and $La_{1-x}Ba_xMnO_3$ thin films. We succeeded to fabricate the LSMO film having good colossal magnetoresistance around the room temperature ($H=0.7$ T).

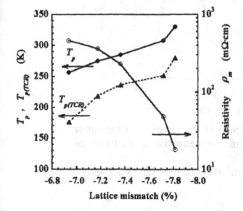

Figure 1. Relationships between the lattice mismatch and the T_p, $T_{p(TCR)}$ and the maximum resistivity ρ_m of the LSMO film

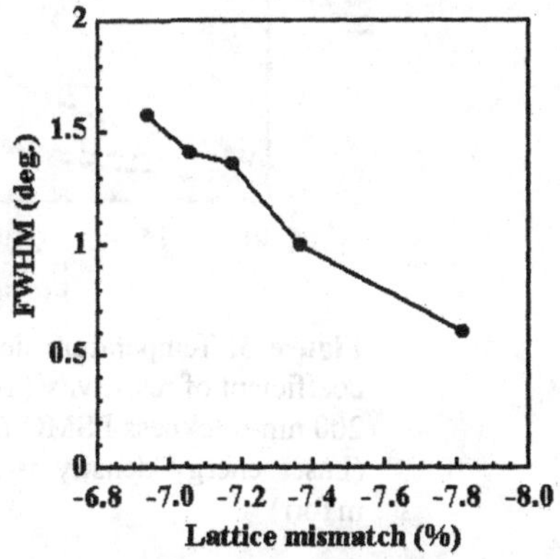

Figure 2. Relationship between the full width at the half maximum of the LSMO(002) rocking curve and lattice mismatch against the MgO substrate.

Table I. Colossal magnetoresistive perovskite thin films.

	T_p(K)	$T_{p(MR)}$(K)	MR (%) (at $T_{p(MR)}$)	MR (%) (at 300K)	H (T)	Process	Substrate	Reference
$La_{1-x}Sr_xMnO_3$								
x=0.20	330	290	15	13	0.7	PLD	MgO	[this paper]
x=0.24	286	210	18	3	0.8	RF sputter	LaAlO₃	[6]
x=0.24	286	235	17	3	0.8	RF sputter	SrTiO₃	[6]
x=0.25	380	330	8	4	0.5	PLD	LaAlO₃	[7]
x=0.33	350	300	28	28	5.0	MOCVD	MgO	[8]
$La_{1-x}Ca_xMnO_3$								
x=0.30	180	172	96	0	5.0	PLD	SrTiO₃	[9]
x=0.33	262	240	28	0	0.5	PLD	LaAlO₃	[7]
x=0.33	170	150	90	10	5.0	MOCVD	MgO	[8]
$La_{1-x}Ba_xMnO$								
x=0.30		315	50	45	5.0	PLD	LaAlO₃	[10]

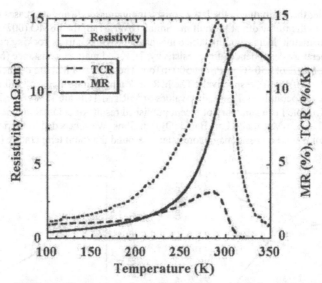

Figure 3. Temperature dependence of the resistivity, temperature coefficient of resistivity (TCR) and magnetoresistance (MR) for the 200 nm-thickness LSMO thin film.
(Laser energy density = 2 J/cm^2 (5 Hz), T_s= 850 °C, P_{O2}=500 mTorr)

PZT / LSMO heterostructures film

Figure 4 and Figure 5 show the AFM surface images of LSMO and PZT layer of the PZT / LSMO / MgO heterostructures, respectively. The mean roughness of the LSMO film is about 1.2 nm so that the good surface morphology (mean roughness of 3.2 nm) of the PZT film is obtained to result in excellent crystallization and ferroelectric properties. Figure 6 is the θ–2θ XRD spectrum of the PZT / LSMO / MgO heterostructures. Both PZT and LSMO films orient c-axis direction. The FWHM of the LSMO (002) and PZT (002) rocking curves were 0.61° and 0.87°, respectively. Figure 7 indicates the φ scan spectrum of the heterostructures. The fourfold symmetries were observed in each layer. The PZT (113) peaks shift by an angle of 45° with respect to the LSMO (103) peaks, which means cube-on-cube epitaxial growth on MgO substrate.

PZT(001)‖LSMO(001)

PZT[100],[010]‖LSMO[100],[010]

The P - E hysteresis loop was measured by the Sawyer-Tower circuit. The Au electrode (Φ=200 μm) was thermally evaporated on the top of PZT film through the metal mask. The Au / PZT / LSMO capacitive structure has the remanent polarization of 22 μC/cm^2 and coercive field of 27 kV/cm at the applied voltage of 7 V$_{p-p}$.

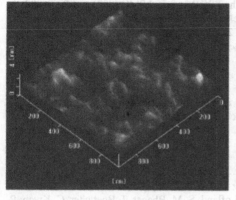

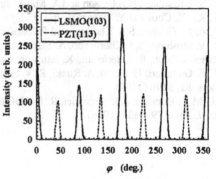

Figure 4. AFM image of the LSMO film (200 nm) on MgO substrate.

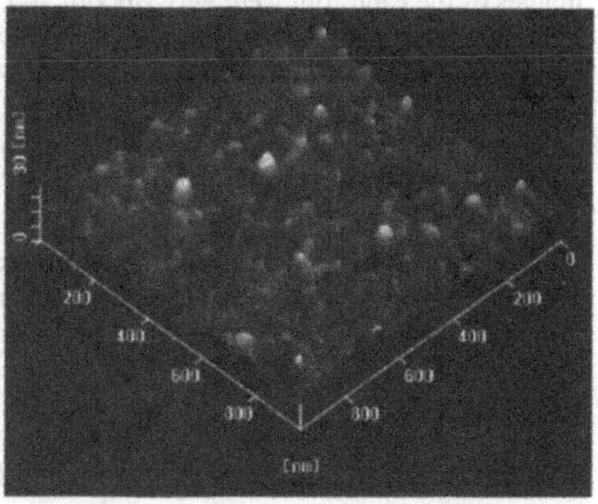

Figure 5. AFM image of the PZT film (400 nm) deposited on the LSMO / MgO.

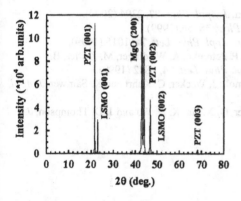

Figure 6. θ–2θ spectrum of the PZT / LSMO / MgO heterostructures with CuK$_\alpha$.

Figure 7. φ scan spectrum of the PZT / LSMO / MgO heterostructures with CuK$_\alpha$.

CONCLUSIONS

We studied the structural and electrical properties of the colossal magnetoresistive LSMO films deposited on MgO (100) substrate under various deposition conditions of T_s and P_{O2}. It was found that high deposition temperature and high oxygen pressure tend to decrease the resistivity, increase T_p and shorten c-axis lattice constant. The large lattice mismatch between LSMO and PZT layer is attributed to increase T_p and decrease the resistivity. The LSMO film deposited at 850°C in 500 mTorr oxygen has T_p of 330 K, TCR of 3% and MR of 15 % (H=0.7 T) at the room temperature. The PZT / LSMO heterostructures grew epitaxially on MgO substrate. The AFM images of the each film were very smooth and the FWHM of the LSMO (002) and PZT (002) rocking curves were 0.61° and 0.87°, respectively. The ferroelectric PZT film had the remanent polarization of 22 $\mu C/cm^2$ and coercive field of 27 kV/cm at the applied voltage of 7 V_{p-p}.

ACKNOWLEDGEMENTS

This work was supported by a Grant-in-Aid for Scientific Research (1998-2000, No. 10045046) by the Ministry of Education, Science, Sports and Culture of Japan.

REFERENCES

1 . P. W. Anderson and H. Hasagawa, *Phys. Rev.* **100**, 675 (1955).

2 . A. Goyal, M. Rajeswari, R.Shreekala, S. E. Lofland, S. M. Bhagat, T. Boettcher, C. Kwon, R. Ramesh and T. Venkatesan, *Appl. Phys. Lett.*, **71**, 2535 (1997).

3 . A. Urushibara, Y. Morimoto, T. Arima, A. Asamitsu, G. Kido and Y. Tokura, *Phys. Rev. B* **51**, 14103 (1995).

4 . F. Mitsugi, T. Ikegami, K. Ebihara,, J. Narayan, A. M. Grishin, *Transactions of the Materials Research Society of Japan* (in print).

5 . S. I. Kharstev, P. Johnsson and A. M. Grishin, *J. Appl. Phys.* **87**, 2394 (2000).

6 . K. –K. Choi and Y. Yamazaki, Jpn. *J. Appl. Phys.* **38**, 56 (1999).

7 . A. M. Grishin, S. I. Khartsev and P. Johnsson, *Appl. Phys. Lett.* **74**, 1015 (1999).

8 . V. Moshnyaga, I. Khoroshun, A. Sidorenko, P. Petrenko, A. Weidinger, M. Zeitler, B. Rauschenbach, R. Tidecks and K. Samwer, *Appl. Phys. Lett.* **74**, 2842 (1999).

9 . E. Gommert, H. Cerva, A. Rucki, R. V. Helmolt, J. Wecker, C. Kuhrt and K. Samwer, *J. Appl. Phys.* **81**, 5496 (1997).

10. M. F. Hundley, J. J. Neumeier, R. H. Heffner, Q. X. Jia, X. D. Wu and J. D. Thompson, *J. Appl. Phys.* **79**, 4535 (1996).

Laser Direct Writing

Mat. Res. Soc. Symp. Vol. 617 © 2000 Materials Research Society

Photo-induced Large Area Growth of Dielectrics with Excimer Lamps

Ian W. Boyd and Jun-Ying Zhang
Electronic and Electrical Engineering, University College London,
Torrington Place, London WC1E 7JE, United Kingdom

ABSTRACT

In this paper, UV-induced large area growth of high dielectric constant (Ta_2O_5, TiO_2 and PZT) and low dielectric constant (polyimide and porous silica) thin films by photo-CVD and sol-gel processing using excimer lamps, as well as the effect of low temperature UV annealing, are discussed. Ellipsometry, Fourier transform infrared spectroscopy (FTIR), X-ray photoelectron spectroscopy (XPS), UV spectrophotometry, atomic force microscope (AFM), capacitance-voltage (C-V) and current-voltage (I-V) measurements have been employed to characterize oxide films grown and indicate them to be high quality layers. Leakage current densities as low as 9.0×10^{-8} $A \cdot cm^{-2}$ and 1.95×10^{-7} $A \cdot cm^{-2}$ at 0.5 MV/cm have been obtained for the as-grown Ta_2O_5 films formed by photo-induced sol-gel processing and photo-CVD, respectively - several orders of magnitude lower than for any other as-grown films prepared by any other technique. A subsequent low temperature (400°C) UV annealing step improves these to 2.0×10^{-9} $A \cdot cm^{-2}$ and 6.4×10^{-9} $A \cdot cm^{-2}$, respectively. These values are essentially identical to those only previously formed for films annealed at temperatures between 600 and 1000°C. PZT thin films have also been deposited at low temperatures by photo-assisted decomposition of a PZT metal-organic sol-gel polymer using the 172 nm excimer lamp. Very low leakage current densities (10^{-7} A/cm^2) can be achieved, which compared with layers grown by conventional thermal processing. Photo-induced deposition of low dielectric constant organic polymers for interlayer dielectrics has highlighted a significant role of photo effects on the curing of polyamic acid films. I-V measurements showed the leakage current density of the irradiated polymer films was over an order of magnitude smaller than has been obtained in the films prepared by thermal processing. Compared with conventional furnace processing, the photo-induced curing of the polyimide provided both reduced processing time and temperature. A new technique of low temperature photo-induced sol-gel process for the growth of low dielectric constant porous silicon dioxide thin films from TEOS sol-gel solutions with a 172 nm excimer lamp has also been successfully demonstrated. The dielectric constant values as low as 1.7 can be achieved at room temperature. The applications investigated so far clearly demonstrate that low cost high power excimer lamp systems can provide an interesting alternative to conventional UV lamps and excimer lasers for industrial large-scale low temperature materials processing.

Keywords: Photo-CVD, excimer lamp, UV annealing, thin Ta_2O_5 film, dielectric constant, sol-gel processing, semiconductor devices

1. Introduction
2.

As dynamic random access memories (DRAMs) are scaled down, the thickness of SiO_2 gate oxide must be correspondingly reduced and thickness control becomes a critical issue. The projected silicon dioxide (SiO_2) thickness by 2012 predicted by the semiconductor industry roadmap will reach atomic dimensions (five silicon atomic layers), or less than one nanometre [1] as indicated in Table 1 [2-3]. As can be seen, for 0.1 μm ultra-large-scale-integrated (ULSI) device technologies, the SiO_2 gate oxide thickness must be scaled below 2.0 nm and become so thin that direct tunneling effects and excessively high electric fields become serious obstacles to reliability as well as fundamental quantum mechanical difficulties. What is required is a thicker layer of a higher dielectric constant material that will have the same or similar effective capacitance when it is put into a device and enable a further decrease in device area (see table 1).

Table 1. The projected gate oxide thickness for CMOS integrated circuits for 1997-2012 [2-3]

Technology timeline	1997	1999	2001	2003	2006	2009	2012
Design rule (μm)	0.25	0.18	0.15	0.13	0.10	0.07	0.05
Wafer diameter (mm)	200	300	300	300	300	450	450
Gate dielectric (nm, ε=3.9) equivalent SiO_2	4-5	3-4	2-3	2-3	1.5-2	<1.5	<1.0
Gate dielectric (nm) Ta_2O_5 thickness (ε=19)	20-25	15-20	10-15	10-15	7.5-10	<7.5	<5

Recently, various high dielectric constant materials have been widely investigated as possible candidates to replace SiO_2 in dynamic random access memories (DRAMs) and shown in table 2. More detailed thermodynamic stability and silicon-compatibility of these dielectrics can be found elsewhere [4-6]. Amongst these dielectrics, much of focus of current high-k research involves PbZr$_x$Ti$_{1-x}$O$_3$ (PZT), Ba$_x$Sr$_{1-x}$TiO$_3$ (BST), Ta$_2$O$_5$, and TiO$_2$. Tantalum pentoxide (Ta$_2$O$_5$) in particular appears to be the most promising and best candidate to replace SiO$_2$ because of its compatibility with ultra-large-scale-integrated processing as well as chemical and thermal stability [7-10]. Various storage capacitor configurations with Ta$_2$O$_5$ dielectric films such as polysilicon/Ta$_2$O$_5$/polysilicon (SIS), metal/Ta$_2$O$_5$/metal (MIM), metal/Ta$_2$O$_5$/polysilicon (MIS or MOS)), etc. have been fabricated to study the nature of Ta$_2$O$_5$ dielectric films [11]. It was also shown that the Ta$_2$O$_5$ capacitor with the TiN/poly-Si top electrode is suitable for 256 Mbit memory devices and has applied to 256 Mbit DRAM fabrication process [9]. Recently, it has been reported the mixed Ta-Ti and Ta-Zr oxides can enhance dielectric constant up to 120 [12-14]. However, fully satisfying the demands of the microelectronics industry for present Ta$_2$O$_5$ films on Si have not yet been fulfilled, although a considerable amount of work on Ta$_2$O$_5$ dielectric films has already been done. Some obstacles need to be overcome before Ta$_2$O$_5$ can be applied to DRAMs technology, especially, the large leakage current in the as-deposited layers. Therefore, approaches aimed at its reduction have received much attention. It was found that the leakage current density could be significantly reduced to acceptable levels by employing of the post-deposition annealing techniques. Various post-deposition annealing techniques including furnace annealing with various gases such as O$_2$, N$_2$O, N$_2$,, NH$_3$, O$_3$ etc., rapid thermal O$_2$ annealing (RTA), ultraviolet generated ozone (UVO), plasma and a range of two-step process as well as several other more complicated methods [15-29], have been proposed to improve the electrical properties of the Ta$_2$O$_5$ films. Since furnace and RTA annealing are carried out at above 700°C the Ta$_2$O$_5$ films are crystallized and consequently grain boundaries lead to an increase in the leakage current [15]. In the case of UVO and plasma annealing, the leakage current can be significantly reduced compared with that obtained by other annealing techniques. Furthermore, Both annealed films remain in the amorphous phase because of low annealing temperatures (<400°C). Table 3 summarizes the properties of Ta$_2$O$_5$ films deposited and annealed by different methods. It is clearly seen that after annealing the leakage current densities significantly reduced to the range 10^{-6}-10^{-8}A/cm^2 from around 10^{-3}A/cm^2 in the as-deposited films. A variety of chemical and physical deposition techniques have been used to deposit Ta$_2$O$_5$ films [8,16-24,30-35]. Future ULSI device architectures preclude the use of high temperature processing for the gate and so the development of low-temperature fabrication routes to these high dielectric constant materials is very important. In order to address this issue of thermal budget, a number of new low temperature growth techniques are now under study including photo-assisted methods. To date most photo-induced processing has been performed with lasers. However, their use is inherently limited by the total photon fluxes available and they are therefore not suited for large area wafer processing. Unlike lasers, lamp sources provide the potential for large area coverage. In particular, the recent development of excimer lamps [36-43], which are capable of producing high power radiation with available wavelengths in the range of 108 nm to 354 nm, has opened up new possibilities for initiating a wide range of large area low temperature photo-induced reactions.

Table 2 Alternate gate dielectrics for use in silicon MOS transistors [4-6]

Material	k	Material	k	Material	k	Material	k
PbZr$_x$Ti$_{1-x}$O$_3$ (PZT)	1000	Ta$_2$O$_5$	25-28	Nd$_2$O$_3$	16-20	MgO	9.8
Ba$_x$Sr$_{1-x}$TiO$_3$ (BST)	500	LaAlO$_3$	25	Y$_2$O$_3$	11-14	Li$_2$O	8.1-8.8
TiO$_2$	100	La$_2$Be$_2$O$_5$	25	Er$_2$O$_3$	12.5-13	MgAl$_2$O$_4$	8.3-8.6
LaScO$_3$	30	ZrO$_2$	22	ZrSiO$_4$	12-13	BeO	6.9-7.7
Y$_2$O$_3$-ZrO$_2$	29.7	La$_2$O$_3$	21	Al$_2$O$_3$	9-12	Ce$_2$O$_3$	7.0

Applications of such sources have already been demonstrated in several areas, including photo-deposition of dielectric [44-47] and metallic [48-55] thin films, photo-oxidation of silicon, germanium and silicon-germanium [56-59], surface modification and polymer etching [60-65], UV annealing [22, 24, 26, 66-67], photo degradation of a variety of pollutants [68-69], as well as large area flat panel displays [70-71] etc. Single- and multi-layered films of silicon oxide, silicon nitride, and silicon oxynitride can be deposited by photo-CVD from different mixtures of silane, oxygen, ammonia, and nitrous oxide gases [44-45, 72-75]. Recent work on the direct photo-oxidation of silicon at low temperature (250°C) has shown the oxidation rate is more than three times greater than that for photo-induced oxidation of silicon using a typical low pressure Hg lamp at 350°C [56,58]. The fixed oxide charge number density (Q/q) has been found to be 4.5 x 10^{10} cm^{-2}, which is comparable to some of the best values reported for thermally grown oxide on Si at a high temperature of 1030°C [76]. In this paper we briefly outline the underlying principles and properties involved in excimer lamp. A novel application towards high and low dielectric constant thin films grown on Si by photo-CVD and sol-gel processing will be reviewed while the effect of low temperature UV annealing is also presented. The physical, optical, electrical and dielectric characteristics of Ta$_2$O$_5$ thin films are summarized and discussed. The applications investigated so far clearly demonstrate that low cost high power excimer lamp systems can provide an interesting alternative to conventional UV lamps and excimer lasers for industrial large-scale low temperature materials processing.

Table 3. Properties of Ta_2O_5 films deposited and annealed by different methods (J is the leakage current density while (a): as-deposited films and (b): annealed films)

Method and annealing conditions	Capacitor structure	k	$J (A/cm^2)$ at 1 MV/cm	Reference
Sputtering Annealed at 850°C	Al/ Ta_2O_5/Pt Al/ Ta_2O_5/p-Si	a) 24 b) 45	a) 10^{-4} b) 10^{-7}	16
CVD at 450°C and RTA annealed at 800°C in O_2, N_2O	Al/ Ta_2O_5/n-Si TiN/Ta_2O_5/n-Si		a)10^{-3} b)10^{-8}	17
Thermal oxidation and O_2 annealed at 800°C	Al/ Ta_2O_5/p-Si	16-25	a) $>10^{-4}$ b)10^{-8}	18
CVD at 400-470°C and O_2 annealed at 700-900°C	Al/ Ta_2O_5/p-Si	19	a) $>10^{-3}$ b)10^{-8}	19-20
Plasma-CVD at 200-600°C and N_2 annealed at 700°C	Al/ Ta_2O_5/p-Si	3. 20.3 4. 19	a) $>10^{-2}$ b)10^{-6}	21
CVD at 400°C and plasma annealed at 400°C	TiN/ Ta_2O_5/n-Si		a)10^{-3} b)10^{-8}	15
Photo-CVD at 250-400°C and UV annealed at 400°C	Al/ Ta_2O_5/n-Si	a) 24 b) 20	a)10^{-3} - 10^{-7} b)10^{-8}	10,22

2.   Characteristics of excimer UV sources

The principle underlying the operation of the excimer lamps relies on the radiative decomposition of excimer (excited dimer) states created by a dielectric barrier discharge (silent discharge) in a rare-gas gas such as Ar_2^* (λ = 126 nm), Kr_2^* (λ = 146 nm), or Xe_2^* (λ = 172 nm) or molecular rare-gas-halide complexes complexes such as ArF* (λ = 193 nm), KrCl* (λ= 222 nm), KrF* (λ = 248 nm), XeCl* (λ = 308 nm). Stevens and Hutton [78] first proposed the concept of excimers, which exist only in the excited state [77] and under normal conditions, do not possess a stable ground state, in 1960. In the last decade, the properties of excimers and kinetics of their formation have been studied extensively by Malinin et al.[79], Volkova et al. [80], Eliasson and Kogelschatz [34-40, 81-82], Neiger et al.[83], and Zhang and Boyd [41-43,85]. In the meantime it has been demonstrated that these excimer UV sources can emit high UV intensities very efficiently [27,56,71], and that large-area UV systems at high power densities, as well as different geometries and wavelengths of excimer lamps have been designed and investigated [39,50,81,86-97].

The mechanism for forming excited rare gases/rare gas halides from ions and electrons begin with dissociative attachment of the electrons to the rare gas/halogen to form positive and negative ions. We use xenon and chlorine as a specific case.

$$e^- \quad + \quad Xe \quad \rightarrow \quad Xe^* \quad + \quad e^- \tag{1}$$
$$e^- \quad + \quad Xe \quad \rightarrow \quad Xe^+ \quad + \quad 2e^- \tag{2}$$
$$e^- \quad + \quad Cl_2 \quad \rightarrow \quad Cl \quad + \quad Cl^- \tag{3}$$

The formation of the rare gas dimer Xe_2^*, occurs through the three-body reaction of excited Xe* with other Xe atom or buffer gas.

$$Xe^* \quad + \quad Xe \quad + \quad M \quad \rightarrow \quad Xe^*_2 \quad + \quad M \tag{4}$$

M is a third collision partner which in many cases can be an atom or molecular of the gases involved or of the buffer gases argon, helium or neon. In most XeCl* exciplexes can be created by the recombination of positive xenon ions and negative chlorine ions (5) or Harpooning reaction (6) which the excited Xe* species directly react with chlorine [98].

$$Xe^+ \quad + \quad Cl^- \quad + \quad M \quad \rightarrow \quad XeCl^* \quad + \quad M \tag{5}$$

$$Xe^* \quad + \quad Cl_2 \quad \rightarrow \quad XeCl^* + \quad Cl \quad\quad\quad\quad\quad (6)$$

These excimer molecules are not very stable and once formed decompose within a few nanoseconds giving up their excitation energy in the form of a VUV or UV photon.

$$Xe^*_2 \quad\quad \rightarrow \quad 2Xe \quad + \quad h\upsilon \text{ (172 nm, VUV radiation)} \quad\quad (7)$$
$$XeCl^* \rightarrow \quad Xe \quad + \quad Cl \quad + \quad h\upsilon \text{ (308 nm, UV radiation)} \quad\quad (8)$$

A large number of different emission spectra of excimers have been investigated from rare-gas excimers, rare-gas halide exciplexes and halogen dimers [36-43,82-83,85]. With excimer lasers only a limited number of wavelengths are available at high power levels. These include $\lambda = 193$ nm (ArF*), $\lambda = 248$ nm (KrF*), $\lambda = 308$ nm (XeCl*), $\lambda = 351$ nm (XeF*), whilst much lower powers can be obtained using F*$_2$ ($\lambda = 157$ nm) and KrCl* ($\lambda = 222$ nm). By contrast, more than 20 different wavelengths can be generated in dielectric barrier discharges, extending their emission bands from the VUV to the visible part of the spectrum [39,60]. Table 4 shows the main peak wavelengths and their corresponding photon energies of rare-gases and rare-gas halides, created by discharge excitation.

Table 4 Peak wavelengths and photon energies of excimer emission bands obtained from various dielectric barrier discharges

Excimer	Wavelength (nm)	Photon energy (eV)	UV range
NeF*	108	11.48	
Ar$_2$*	126	9.84	
Kr$_2$*	146	8.49	
F$_2$*	158	7.85	
ArBr*	165	7.52	VUV
Xe$_2$*	172	7.21	
ArCl*	175	7.08	
KrI*	190	6.49	
ArF*	193	6.42	
KrBr*	207	5.99	
KrCl*	222	5.58	
KrF*	248	5.01	UV-C
XeI*	253	4.91	
Cl$_2$*	259	4.79	
XeBr*	283	4.41	
Br$_2$*	289	4.29	UV-B
XeCl*	308	4.03	
I$_2$*	342	3.63	UV-A
XeF*	351	3.53	

Different geometrical configurations of excimer lamps can be designed and fabricated, such as cylindrical, with UV radiating to the outside or the inside, or planar, with UV radiating to one or both sides and windowless for VUV radiation [49-50,60]. Cylindrical lamps were used in most of our work presented here. A typical lamp consisted of two concentric quartz tubes, outer and inner metallic electrodes, an external high voltage generator, and cooling water. The measured energy conversion efficiencies (UV output/electrical input) for these lamps can be as high as 22.5% [70]. Unlike the classical discharge, in the silent discharge microdischarges generally occur. In the work described here the discharge gap was filled with either a rare-gas or a rare gas-halogen gas mixture, and the photons were emitted through a quartz and outer electrode which was transparent to the radiation generated. An alternating voltage of typically a few kV amplitude is adequate to run the discharge. The frequency of the applied voltage can vary over a wide range from 50 Hz to several MHz. The properties of these microdischarges have been extensively studied [36,39,41-42,82].

A large area system has been designed and constructed with an array of parallel excimer lamps, capable of supplying uniform radiation over a 500 cm^2 area. A theoretical model of the UV intensity distribution of such a system has been developed, which assumes the excimer lamp to be a cylindrical source taking into account the fact that it consists of many microdischarges spread over the surface of the cylinder. The UV intensity distribution for a three excimer lamp system measured by using photodiode, photomultiplier, and chemical actinometric methods [41-42,99-100]. The uniformity of the UV intensity is seen to be within ±4% over a 250 cm^2 area. The measurements are in excellent agreement with the modeling [70].

The average life of conventional lamps is governed by the decrease of luminous flux caused by the unavoidable deposition of evaporated electrode material (most usually tungsten) on the inner wall of the envelope because the electrodes are directly in contact with the discharge gas and plasma as previously mentioned. After about a 5% weight loss from the electrode, breakage of the lamp may occur. For most mercury lamps lifetime is between 500-2000 hours, depending on the precise type of lamp. For certain specialized purposes (e.g. photographic lighting) the electrode may reach a temperature as high as 3300K (melting point of tungsten: 3655K), so that the lifetime is reduced to only a few hours. On the contrary, for excimer UV lamps the electrodes are not in direct contact with the discharge gases, and thus avoid any corrosion during the discharge process, thereby providing the excimer lamp with a long lifetime. It was found that the 100% level of the original UV intensity for the 222 nm and 308 nm lamps was still output up to 4000 hours operating time [101].

It can thus be seen that the use of excimers for the generation of UV radiation offers several important advantages over other lamp systems: 1) high intensity at the defined wavelength, 2) no-self absorption, 3) a long lifetime and no contamination of the excimer gas nor electrode corrosion, 4) non-toxic materials are used (e.g. no Hg) and thus inherently there is minimal environmental problem, 5) flexibility to design in different geometries, and 6) potential for scalability to large areas. Because of these unique properties, and additionally their simplicity of construction, large emission area with high-energy VUV photons, low cost, and availability of different wavelengths, excimer sources are an attractive alternative to conventional UV lamps and lasers for large-area industrial applications. Several of their novel applications towards dielectric film formation will be reviewed in the following sections.

3. Photo-induced processing

Figure 1 shows a schematic diagram of large area excimer lamps system, in which comprised a set of two stainless steel chambers separated by a MgF$_2$ window transparent to the VUV radiation. The lamp chamber consisted of an array of parallel cylindrical lamp tubes, details of which are shown on the right side of Fig. 1. The lamps contained either pure xenon (Xe$^*_2$, λ=172nm) or a mixture of krypton and chlorine (KrCl*, λ=222nm). Excimer UV radiation was generated in the top chamber traversed a low pressure gas phase mixture contained within the bottom chamber, and impinged upon the sample with output power, in the range of 10-200 mW/cm^2, determined using actinometric techniques [99-100].

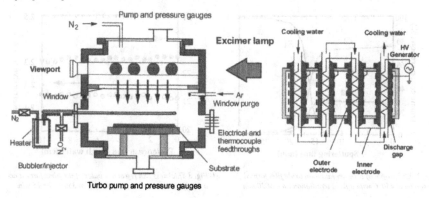

Fig. 1   Schematic diagram of photo-induced process incorporating of an array of excimer UV lamps.

N-type single crystal (100) orientation silicon (2-4 Ω·cm resistivity) wafers were used as substrates, which cleaned using a standard RCA, clean prior to use. The substrate temperature was maintained between 150 and 450°C and measured with a thermocouple attached to the heater stage. The tantalum metalorganic precursors, namely tantalum tetraethoxy dimethylaminoethoxide (Ta(OEt)$_4$(DMAE)) and tantalum ethoxide, were vaporized at temperatures between 100-130°C in a bubbler/injector and then transported into the reaction chamber by an N$_2$ carrier gas. A full description of this reactor is published elsewhere [26]. The processing chamber could be evacuated to 10^{-6} mbar by a turbomolecular pump and filled with the appropriate gas mixture for specific processing applications. UV annealing was performed on the films at different exposure times at temperatures between 350°C and 400°C for a fixed pressure of 1000 mbar in high purity oxygen (99.999%) by using 172 nm excimer lamps.

The chemical compositions in the films were determined by X-ray photoelectron spectroscopy (XPS) using a VG ESCALAB 220i XL. The depth profiles in the samples of Ta, O, and Si were determined using Ar$^+$ ions (3 keV) with a current of 0.8 μA at an argon pressure of 10^{-7} torr. The ratio of oxygen to tantalum in the films was calculated by peak deconvolution of the XPS curves. The thickness and refractive index of the films were determined using a Rudolph AutoEL II ellipsometer while the structure and optical properties of the layers grown on Si wafer and quartz were measured using a Fourier transform infrared (FTIR) spectrometer (Paragon 1000, Perkin Elmer) and UV spectrophotometry, respectively. Surface morphology characterization was observed by using a Topometrix AFM operating in contact mode to determine surface defects. The electrical properties of the films were measured at a frequency of 1 MHz by HP4140 and HP4275 semiconductor systems (I-V, C-V) on Al/Ta$_2$O$_5$/Si capacitor test structures with an evaporated Al top contact of area 8 x 10^{-4} cm^2 through a metal contact mask.

4. Photo-induced large area growth of dielectrics with excimer lamps
4.1. Large area growth of Ta$_2$O$_5$ by photo-induced sol-gel processing

To form a sol-gel solution, tantalum ethoxide (Ta(OC$_2$H$_5$)$_5$) was dissolved in ethanol with a small quantity of water and hydrochloric acid in ethanol. In the sol-gel process, the reaction involves two simultaneous chemical processes, hydrolysis and polymerization. The alkoxide hydrolysis and polymerization reactions occurred over several hours, during which the colloidal particles and condensing metal species linked together to become a three dimensional network, i.e., a slow polymerization of the organic compounds took place, leading to gelation. From this sol-gel solution, films of 10-300 nm thickness were prepared on Si (100) substrates by the spin-on method and then irradiated for various times at different temperatures, to form tantalum oxide [47].

An XPS profile of a 14 nm tantalum oxide film growth on silicon at 450°C at a fixed irradiation time of 20 min is shown in Fig. 2. The concentration of tantalum and oxygen is fairly constant throughout the Ta$_2$O$_5$ film. No carbon was observed although it is always present in films obtained by plasma-CVD and thermal-CVD [102]. The atomic ratio of O/Ta was between 2.4 - 2.6 which is very close to the stoichiometric ratio of 2.5 for Ta$_2$O$_5$, and higher than the 2.2 ratio for films photo-irradiated by a low pressure mercury lamp and the 2.1 ratio reported for both heat-treated layers and photo-deposited films [102]. The average refractive index obtained at temperatures above 350°C was 2.15±0.05, which is close to the bulk Ta$_2$O$_5$ value of 2.2.

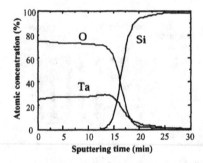

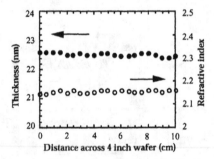

Fig. 2 XPS profile of a 14 nm tantalum oxide film formed on silicon at 450°C with a fixed irradiation time of 20 min

Fig. 3 Thickness and refractive index of the films formed across a 4 inch silicon wafer by irradiating at 350°C for 15 min.

Figure 3 shows the thickness and refractive index of layers formed on a 4 inch silicon wafer by irradiating spin-coated films at a temperature of 350°C for an exposure time of 15 min. As can been seen, very uniform films (about 22.6 nm) were achieved with the total variation in thickness across the 4-inch wafer being within ± 0.2 nm. The mean refractive index measured (2.15) was similar to that usually obtained for plasma-deposited Ta$_2$O$_5$ films [29]. XRD revealed the structure of films formed to be amorphous.

Metal oxide semiconductor (MOS) capacitors have been fabricated using 20 nm thick as grown Ta$_2$O$_5$ layers formed by this process. Table 5 shows a comparison of the electrical properties in our films formed at different temperatures. It can be seen that the fixed oxide charge density decreases with increasing temperature. The fixed oxide charge density changed from 4.0 x 10^{11} cm^{-2} at 150°C to 1.0 x 10^{11} cm^{-2} at 400°C, which is similar to those obtained in the films prepared by plasma-CVD processing [103]. The I-V characteristics of the MOS capacitors also showed that the leakage current density reduced dramatically in films grown at 400°C (see Table 5), indicating a more ideal reaction between oxygen and tantalum species at the higher temperatures.

Table 5 Comparison of the electrical properties of the as grown Ta₂O₅ films at different temperatures

Temperature (°C)	Fixed charge density (cm^{-2})	Leakage current density at 0.5 MV/cm (A/cm^2)
150	4.0×10^{11}	1.9×10^{-5}
250	2.6×10^{11}	9.2×10^{-6}
400	1.0×10^{11}	9.0×10^{-8}

A leakage current density as low as 9.0×10^{-8} A·cm^{-2} at 0.5 MV/cm can be achieved, which is over 2 orders magnitude lower than those obtained in as grown films prepared by plasma-CVD method (see table 6) [31]. A subsequent low temperature (400°C) annealing in UV improves this to 2.0×10^{-9} A·cm^{-2} at 0.5 MV/cm (table 4). These values are comparable to those only previously obtained for films annealed at temperatures between 600° and 1000°C [21,31].

Table 6 Comparison of the leakage current densities at 0.5 MV/cm (A/cm^2) in Ta₂O₅ films obtained by different methods [31].

	Plasma-CVD		Our work	
as-deposited	annealing at 700-800°C	as-deposited	annealing at 400 °C
10^{-6}	$10^{-8} - 10^{-9}$	9.0×10^{-8}	2.0×10^{-9}

4.2. Thin Ta₂O₅ film grown by photo-CVD

In photo-induced CVD of Ta₂O₅ from tantalum ethoxide (Ta(OC₂H₅)₅) and nitrous oxide, the primary photochemistry of the N₂O involves the following reaction:

$$N_2O \; + \; h\upsilon \; \rightarrow \; O(^1D) \; + \; N_2 \, (X^1 \cdot {}^+_g) \tag{9}$$

The active oxygen species O (^{1}D) subsequently react with the Ta(OC₂H₅)₅ causing its dissociation through a series of reactions leading to Ta₂O₅ deposition on the substrate surface.

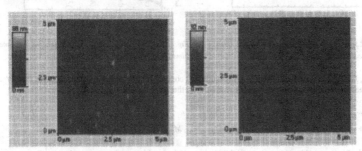

Fig. 4 AFM image of Ta₂O₅ films deposited at 200°C (left) and 350°C (right)

Figure 4 shows on AFM image of the surface features of Ta₂O₅ films deposited at 200°C (left of Fig. 4) and 350°C (right of Fig. 4).As can be seen, AFM image of the films deposited at 200°C revealed the presence of droplets varying in size from 10-500 nm which were not observed for substrate temperature over 300°C. A uniform structure down to a nanostructure scale with particle sizes of about 20 nm was observed by AFM at a temperature of 350°C.

XPS analysis showed that the atomic ratio of O/Ta, of about 2.4, is very close to the stoichiometric ratio of 2.5 for Ta₂O₅ [10]. Fig. 5 shows the evolution of the FTIR spectra in the 400-4000 cm^{-1} range for an as-deposited film, which was then annealed using a 172 nm lamp. The spectra exhibit one dominant peak centered around 650 cm^{-1} and two shoulder-like peaks near 530 cm^{-1} and in the 800-1000 cm^{-1} region agreeing with observations in previous work [24,104]. The peaks at 530 cm^{-1} and 650 cm^{-1} are assigned to the absorption of Ta-O-Ta and Ta-O stretching vibrational modes, characteristic of tantalum pentoxide [24,104]. The weak absorption band at 800-1000 cm^{-1} was attributed to the presence of suboxides [24]. It is clearly seen that UV annealing can significantly reduce this suboxide absorption and completely remove the H_2O and OH groups at 3400 cm^{-1}.

UV spectral measurements showed that the average transmittance of 90% in the visible region of the spectrum for films deposited at temperatures between 250-400°C, which is characteristic of very high quality Ta_2O_5 films [26].

The I-V characteristics of the MOS capacitors also show that after UV annealing, the leakage current density is reduced dramatically as shown in Fig. 6 where it can be seen that it decreases with increased annealing time. After 1 h annealing, leakage current densities as low as 6.4×10^{-9} A/cm^2 at 0.5 MV/cm are achieved. This is two orders of magnitude lower than for as-deposited layers (1.95×10^{-7} A/cm^2) and comparable to values only previously achieved for films annealed at high temperatures (600-900°C) [11,18,21]. Several effects could cause the reduction of leakage current, but here we consider three of the most likely contributors. First, the active oxygen species formed by the 172 nm light can assist in reducing or removing any suboxides present leading to improved stoichiometry as shown in the FTIR (Fig. 5), where suboxides in the as-deposited films were clearly removed by UV annealing. This effect has already been reported for layers grown by pulsed laser deposition [24]. Second, it is known that the active oxygen species created can decrease the density of defects and oxygen vacancies in the as-deposited film [10]. Additionally, it has been reported and confirmed by XPS and FTIR that SiO_2 layers can be formed at the Ta_2O_5/Si interface and on the surface of Ta_2O_5, by the reaction between the active oxygen and Si during annealing, leading to improved interfacial quality [18,105]. All of these could in some measure lead to the reduction of leakage current density in our layers although it is not clear at present which of these dominates.

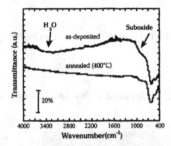

Fig. 5 FTIR spectra for Ta_2O_5 films deposited at 350°C and UV annealed at 400°C.

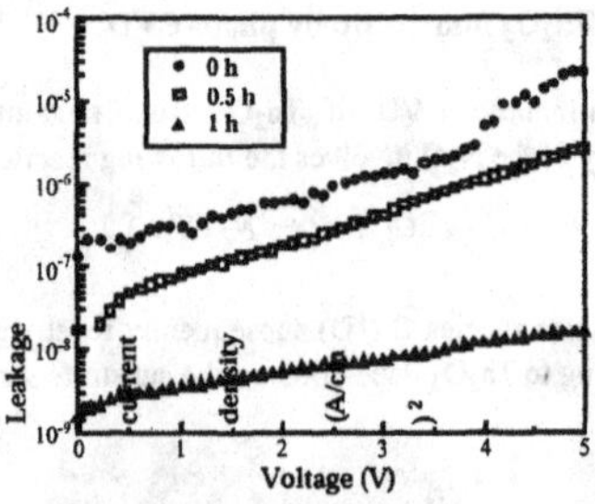

Fig. 6 Leakage current density of as-deposited and annealed films for different times

From XPS, FTIR, and electrical measurement results, we conclude that a possible mechanism for the UV annealing effect is attributed to the reactive oxygen species produced by the UV irradiation as follows:

$$O_2 \quad + \quad h\upsilon \,(\lambda = 172\,nm) \quad \rightarrow \quad O\,(^3p) \quad + \quad O\,(^1D) \tag{10}$$

$$O_2 \quad + \quad O\,(^3p) \quad + \quad M \quad \rightarrow \quad O_3 \quad + \quad M \qquad (M \text{ is a third body}) \tag{11}$$

The ozone is decomposed by absorption of VUV light at 172 nm which produces excited state ^1D oxygen atoms.

$$O_3 \quad + \quad h\upsilon \,(\lambda = 172\,nm) \quad \rightarrow \quad O_2 \quad + \quad O\,(^1D) \tag{12}$$

The active oxygen species formed react with silicon which diffuses from the Si substrate to Ta_2O_5 surface leading to the formation SiO_2 at the surface of the Ta_2O_5 films. On the other hand, the active oxygen species can diffuse through the thin SiO_2 oxide, and react with Ta suboxides to create Ta_2O_5, and possibly also remove certain defects and oxygen vacancies present in the Ta_2O_5.

4.3. Other high dielectric constant materials (TiO_2 and PZT) formed by photo-induced process

Photo-induced sol-gel processing of other high dielectric constant materials such as TiO_2 and PZT using the excimer UV sources has also been demonstrated [106-107]. Single and multiple layer TiO_2 films have been successfully prepared at low temperatures by photo-induced sol-gel processing using a 172 nm excimer lamp [94]. Refractive index values ranging from 2 to 2.4 were measured for multilayers irradiated for 10 min (see fig. 7). These values compare favorably with the value of 2.58 for the bulk material. The films formed showed good optical properties with transmittance values between 85% and 90% in the visible range of the spectrum.

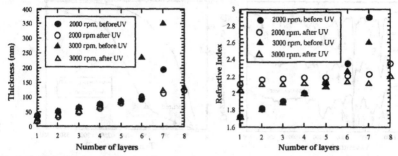

Fig. 7 *Thickness and refractive index of TiO₂ films before and after irradiation at 172 nm for different layers.*

PZT thin films have also been deposited at low temperatures by photo-assisted decomposition of a PZT metal-organic sol-gel polymer using the 172 nm excimer lamp [107]. Very low leakage current densities (10^{-7} A/cm^2) can be achieved, which compared with layers grown by conventional thermal processing. This particular photo-induced approach not only enables reduced temperatures and processing times to be used but also provides good electrical properties without the need for high temperature annealing.

4.4. Low dielectric constant materials (polyimide and porous silica)

The performance of ultra-large scale integrated (ULSI) devices becomes crucial at the metal interconnect level when the feature sizes are reduced to low sub-micron dimensions. The gain in device speed at an MOS device gate is offset by the propagation delay at the metal interconnects due to the increased RC (resistance and capacitance) time constant. This RC time delay can be reduced either with the incorporation of low permittivity dielectric materials and/or high conductivity metals. Polymeric films are one of the most promising groups of low dielectric constant materials which may eventually replace the widely used SiO₂ as an interlayer dielectric to shorten RC time delays, reduce "cross-talk" between metal lines and decrease power consumption at high signal frequencies [108]. Polyimides are particularly attractive not only because of their low dielectric constant, but also their ease of application and patterning and high thermal stability [109]. Recent work on photo-induced deposition of low dielectric constant organic polymers for interlayer dielectrics has highlighted a significant role of photo effects on the curing of polyamic acid films [110]. Compared with conventional furnace processing, the photo-induced curing of the polyimide provided both reduced processing time and temperature. In particular, I-V measurements showed that the leakage current density of the irradiated polymer was over an order of magnitude smaller than has been obtained in layers prepared by thermal processing [110].

Figure 8 presents the FTIR spectra, in the 600-2000 cm^{-1} range for films after the initial prebake and after a 150°C cure for 20 min with the lamp off (conventional thermal curing) and under otherwise identical conditions, but with the additional UV irradiation. Figure 8a shows the characteristic bands of the carboxyl (–COOH) absorption at 1723 cm^{-1}, amide (–CONH–) groups at 1659 and 1546 cm^{-1}, and the amide stretching mode in polyamic acid at 1410 cm^{-1}. After the UV curing (Fig. 8c) all these bands completely disappear. Simultaneously, a typical doublet of a carbonyl group corresponding to an imide moiety appears at 1778 and 1726 cm^{-1} together with the imide C-N absorption band at 1379 cm^{-1}. The small absorption at 728 cm^{-1} has been attributed to deformation of the imide ring or the imide carbonyl groups [111]. The bands corresponding to polyimide at 1778, 1726 and 1379 cm^{-1} for the thermally cured polyamic acid (Fig. 8b) are significantly smaller than those obtained by UV curing. Also the bands related to polyamic acid at 1659, 1546, 1410 cm^{-1} decreased but did not completely disappear. These results indicate that the polyamic acid film is completely transformed to polyimide by the UV curing step at 150°C, whilst the thermally cured sample is only partly transformed.

The degree of imidization at different temperatures for both the UV curing and purely thermal curing steps is shown in Fig. 9. The degree of imidization was calculated by comparing the 1375 cm^{-1} imide band and the 1500 cm^{-1} aromatic band intensities, which are known to give precise internally consistent measurements [111]. At lower curing temperatures (i.e. <150°C) the imidization characteristics of the two curing methods are markedly different. For UV curing, the films start to imidize very significantly, whilst the imidization of the films is very slow for thermal curing. The degree of imidization is 85% for UV curing at 150°C, whilst it is less than 20% for the thermal process.

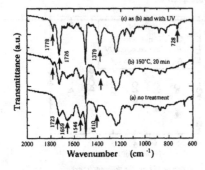 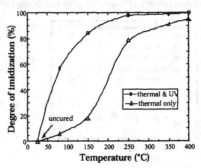

Fig. 8 FTIR spectra of polyamic acid films after a 150°C cure for 20 min with and without UV irradiation, a) initial prebake; b) 150°C, 20 min; c) 150°C, irradiated 20 min by 172 nm lamp.

Fig. 9 Degree of imidization of polyamic acid film as a function of curing temperature with and without irradiation

Very recently, low dielectric porous silica films have been successfully grown from TEOS sol-gel solutions at low temperatures using an excimer lamp [112]. Dielectric constant values as low as 1.7 can be achieved in films prepared at room temperature. These results indicate that this low temperature photo-induced sol-gel technique is very promising for the preparation of low-k dielectric polymer and porous silica films or other interlayer dielectrics in future ULSI multilevel interconnections.

5. Conclusions

Excimer UV lamps with their unique properties, providing high intensity narrow-band radiation over large-areas at a number of different wavelengths, are an interesting alternative to conventional UV lamps and lasers for industrial large-scale low temperature processes. The application of these sources towards low temperature deposition of high dielectric constant (Ta_2O_5, TiO_2 and PZT) and low dielectric constant (polyimide and porous silica) materials has been successfully demonstrated. Very uniform Ta_2O_5 films were achieved with the total variation in thickness across the 4-inch wafer being within ± 0.2 nm. The leakage current density of the irradiated dielectric films was over several orders of magnitude smaller than that obtained by thermal processing. UV annealing can significantly reduce the leakage current density of Ta_2O_5 thin films incorporated in simple MOS capacitor structures. The active oxygen species, produced by 172 nm radiation, are considered to play an important role in the reduction of the leakage current density since they can reduce the density of defects and oxygen vacancies, remove any suboxides and impurities present in the films. This photo-induced process also enables reduced temperatures and times to be used. Therefore, UV-induced low temperature deposition of thin films offers a very effective method for fabrication of high quality thin films for industrial applications in electronic and optical manufacturing.

Acknowledgement:
The authors would like to thank Dr. U. Kogelschatz (ABB, Corporate Research, Switzerland) for many stimulating discussions. This work was partly supported by EPSRC (grant No. GR/190909).

References
1. D.A. Muller, T. Sorsch, S. Moccio, F.H. Baumann, K. Evans-Lutterodt and G. Timp, Nature, 399 (1999) 758.
2. Semiconductor industry Association The National Technology Roadmap for Semicond. 71-78 (Sematech Austin, 1997)
3. M. Schulz, Nature, 399 (1999) 729.
4. C.A. Billman, P.H. Tan, K.J. Hubbard, and D.G. Schlom, Mat. Res. Soc. Symp. Proc. 567 (1999) 409.
5. Q.X. Jia, X.D. Wu, S.R. Foltyn, and P. Tiwari, Appl. Phys. Lett. 66 (1995) 2197.
6. R. Singh, S. Alamgir, and R. Sharangpani, Appl. Phys. Lett. 67 (1995) 3939.
7. S. Tanimoto, M. Matsui, K. Kamisako, K. Kuroiwa, and Y. Tarui, J. Electrochem. Soc. 139 (1992) 320.

8. Y. Nishimura, K. Tokunaga and M. Tsuji, Thin Solid Films 226 (1993) 144.
9. KW Kwon, CS Kang, SO Park, HK Kang, ST Ahn, IEEE Trans Electron Devices 43 (1996) 919.
10. J.-Y. Zhang, B. Lim, V. Dusastre, and I.W. Boyd, Appl. Phys. Lett. 73 (1998) 2299.
11. H. Shinriki, T. Kisu, Y. Nishioka, Y. Kawamoto, and K. Mukai, IEEE Trans. Electron Devices 37 (1990) 1939.
12. R.F. Cava, W.F. Peck Jr and J.J. Krajewski, Nature, 377 (1995) 215.
13. A. Cappellani, J.L. Keddie, N.P. Barradas and S.M. Jackson, Solid-State Electronics, 43 (1999) 1095.
14. R.J. Cava and J.J. Krajewski, J. Appl. Phys. 83 (1998) 1613.
15. S. Kamiyama, H. Suzuki, H. Watanabe, A. Sakai, H. Kimura, and J. Mizuki, J. Electrochem Soc., 141 (1994) 1246.
16. A. Pignolet, G. M. Rao, S.B. Krupanidhi, Thin Solid Films, 258 (1995) 230.
17. S.C. Sun and T.F. Chen, IEEE Electron Device letters 17 (1996) 355.
18. S.W. Park, Y.K. Baek, J.Y. Lee, C.O. Park and H.B. Im, J. Electronic. Mater. 21 (1992) 635.
19. S. Zaima, T. Furuta, Y. Koide, and Y. Yasuda, , J. Electrochem. Soc. 137 (1992) 2876.
20. S. Kamiyama, P. Lesaicherre, H. Suzuki, I. Nishiyama and A. Ishitani, J. Electrochem. Soc. 140 (1993) 1617.
21. J.L. Autran, P. Paillet, J.L. Leray, and R.A.B. Devine, Sensors and Actuators A51 (1995) 5.
22. J.-Y. Zhang, V. Dusastre, D.E. Williams and I.W. Boyd, J. Phys. D: Appl. Phys.32 (1999) L1.
23. H. Shinriki and M. Nakata, IEEE Trans. Electron. Dev. 38 (1991) 455.
24. J.-Y. Zhang, Q. Fang, and I.W. Boyd, Appl. Surf. Sci. 138-139 (1999) 320.
25. R.A.B. Devine, Appl. Phys. Lett. 68 (1996) 1924.
26. J.-Y. Zhang, B. Lim, and I.W. Boyd, Thin Solid Films 336 (1998) 340.
27. C. Isobe and M. Saitoh, Appl. Phys. Lett. 56 (1990) 907.
28. J.-Y. Zhang and I.W. Boyd, J. of Mater. Sci. Lett. 17 (1998) 1507.
29. P.A. Murawala, M. Sawai, T. Tatsuta, O. Tsuji, S. Fujita, and S. Fujita, Jpn. J. Appl. Phys. 32 (1993) 368.
30. H.O. Sankur and W. Gunning, Appl. Opt. 28 (1989) 2806.
31. I.L. Kim, J.S. Kim, O.S. Kwon, S.T. Ahn, J.S. Chun and W.J. Lee, J. Electron. Mater. 24 (1995) 1435.
32. D. Laviale, J.C. Oberlin, and R.A.B. Devine, Appl. Phys. Lett. 65 (1994) 2021.
33. M. Matsui, S. Oka, K. Yamagishi, K. Kuroiwa, and Y. Tarui, Jpn. J. Appl. Phys. 27 (1988) 506.
34. S. Oshio, M. Yamamoto, J. Kuwata, and T. Matsuoka, J. Appl. Phys. 71 (1992) 3471.
35. T. Aoyama, S. Yamazaki, and K. Imai, J. Electrochem. Soc. 145 (1998) 2961.
36. B. Eliasson and U. Kogelschatz, Appl. Phys. B 46 (1988) 299.
37. B. Eliasson and U. Kogelschatz, Proc. 40 Ann. Gas. Electron. Conf. (GEC 87), Atlanta 1987, p.174.
38. B. Gellert, B. Eliasson and U. Kogelschatz, Proc. 5 Int. Symp. on the Science & Technology of Light Sources (LS:5), York 1989, p.155 and 181.
39. U. Kogelschatz, Pure & Appl. Chem. 62 (1990) 1667.
40. U. Kogelschatz, Appl. Surf. Sci. 54 (1992) 410.
41. J.-Y. Zhang and I.W. Boyd, J. Appl. Phys. 80 (1996) 633.
42. J.-Y. Zhang and I.W. Boyd, J. Appl. Phys. 84 (1998) 1174.
43. I.W. Boyd and J.-Y. Zhang, Nucl. Instrum. Meth. Phys. Res. B121 (1997) 349.
44. P. Bergonzo and I.W. Boyd, J. Appl. Phys. 76 (7) (1994) 4372.
45. P. Bergonzo and I.W. Boyd, Appl. Phys. Lett. 63 (1993) 1757.
46. J.-Y. Zhang, L.-J. Bie, and I.W. Boyd, Jpn. J. Appl. Phys. 37 (1998) L27.
47. J.-Y. Zhang, B.-J. Bie, V. Dusastre and I.W. Boyd, Thin Solid Films 318 (1998) 252.
48. H. Esrom, J. Demny, and U. Kogelschatz, Chemtronics 4 (1989) 202.
49. H. Esrom and U. Kogelschatz, Appl. Surf. Sci. 46 (1990) 158.
50. H. Esrom and U. Kogelschatz, Appl. Surf. Sci. 54 (1992) 440.
51. J.-Y. Zhang, Qi Fang, S.L. King and Ian W. Boyd, Appl. Surf. Sci. 109/110 (1997) 487.
52. J.-Y. Zhang, H. Esrom, and I.W. Boyd, Appl. Surf. Sci. 96-98 (1996) 399.
53. J.-Y. Zhang and I.W. Boyd, J. Mat. Sci. Lett. 16 (1997) 996.
54. J.-Y. Zhang and I.W. Boyd, Appl. Phys. A 65 (1997) 379.
55. J.-Y. Zhang and I.W. Boyd, Thin Solid Films, 318 (1998) 234.
56. J.-Y. Zhang and I.W. Boyd, Electronics Letters, 32 (1996) 2097.
57. V. Cracium, B. Hutten, D.E. Williams and I.W. Boyd, Electronics Letters 34 (1998) 71
58. J.-Y. Zhang and I.W. Boyd, Appl. Phys. Lett. 71 (1997) 2964.
59. V. Craciun, J-Y. Zhang and I.W. Boyd, NATO Fund. Aspects of Ultrathin Dielectrics on Si-based Dev. 1997, pp461.
60. H. Esrom and U. Kogelschatz, Thin solid films, 218 (1992) 231.
61. J.-Y. Zhang, Thesis, Karlsruhe University, Germany, 1993.
62. H. Esrom, J.-Y. Zhang, and U. Kogelschatz, Mat. Res. Symp. Proc. 236 (1992) 39.
63. J.-Y. Zhang, H. Esrom, U. Kogelschatz and G. Emig, Appl. Surf. Sci. 69 (1993) 299.
64. J.-Y. Zhang, H. Esrom, U. Kogelschatz, and G. Emig, J. of Adhesion Sci. and Technol. 8 (1994) 1179 .
65. J.-Y. Zhang, H. Esrom, and I.W. Boyd, Surface and Interface Analysis 24 (1996) 718.

66. V. Craciun, I.W. Boyd, D. Craciun, P. Andreazza and J. Perriere, J. Appl. Phys. 85 (1999) 8841.
67. V. Craciun, D. Craciun, P. Andreazza, J. Perriere and I.W. Boyd, Appl. Surf. Sci. 139 (1999) 587.
68. U. Kogelschatz, NATO Advanced Research Workshop on Non-thermal Plasma Techniques for Pollution Control, Cambridge University, UK, September 21-25, 1992.
69. R.S. Nohr and J.G. MacDonald, U. Kogelschatz, G. Mark, H.-P. Schuchmann and C. von Sonntag, J. Photochem. Photobiol. A: Chem. 79 (1994) 141.
70. J.-Y. Zhang and I.W. Boyd, Mat. Res. Symp. Proc. 471 (1997) 53.
71. T. Urakabe, S. Harada, T. Saikatsu and M. Karino, Sci. and Tech. of light Sources (LS7) Kyoto, 1995, Eds: R. Italani nd S. Kamiya, pp159.
72. P. Bergonzo, U. Kogelschatz, and I.W. Boyd, Appl. Surf. Sci. 69 (1993) 393.
73. P. Bergonzo, U. Kogelschatz, and I.W. Boyd, SPIE, Vol 2045 (1994) 174.
74. P. Bergonzo and I.W. Boyd, Electronics Letters, 30 (1994) 606.
75. P. Bergonzo and I.W. Boyd, Microelectronic Engineering 25 (1994) 345.
76. G. Eftekhari, J. Electrochem. Soc. 140 (1993) 787.
77. B. Gellert, U. Kogelschatz, Appl. Phys. B52 (1991) 14 .
78. B. Stevens and E. Hutton, Nature, 186 (1960) 1045 .
79. A.N. Malinin, A.K. Shuaibov and V.S. Shevera, J. Appl. Spectrosc., 32 (1980) 313 .
80. G.A. Volkova, N.N. Kirillova, E.N. Pavlovskaya and A.V. Yakovleva, J. Appl. Spectrosc. 41 (1984)1194.
81. B. Eliasson and B. Gellert, J. Appl. Phys. 68 (1990) 2026 .
82. B. Eliasson, M. Hirth and U. Kogelschatz, J. Phys D: Appl. Phys. 20 (1987) 1421.
83. M. Neiger, V. Schorpp and K. Stockwald, Proc. 41. Ann. Gaseous Electron. Conf. (GEC 88), Minneapolis p.74, 1988.
84. I.W. Boyd and J.-Y. Zhang, Mat. Res. Symp. Proc. 470 (1997) 343.
85. J.D. Ametepe, J. Diggs, D.M. Manos and M.J. Kelley, J. Appl. Phys. 85 (1999) 7505.
86. P. Patel, I.W. Boyd, Appl. Surf. Sci. 46 (1990) 352 .
87. F. Kessler and G.H. Bauer, Appl. Surf. Sci. 54 (1992) 430.
88. F. Kessler, H.D. Mohring, G.H. Bauer, Proc. of the 9th Conf. on Plasma Chem. 3 (1989) 1383.
89. C. Manfredotti, F. Fizotti, M. Boero, G. Piatti, Appl. Surf. Sci. 69 (1993) 127.
90. B. Bollanti, G. Clementi, P.D. Lazzaro, F. Flora, G. Giordano, T. Letardi, F. Muzzi, G. Schina and C.E. Zheng, IEEE Transactions on Plasma Science 27 (1999) 211.
91. A.K. Shuaibov, L.L. Shimon and I.V. Shevera, Instr. and Experimental Tech. 41 (1998) 427.
92. P.N. Barnes and M.J. Kushner, J. Appl. Phys. 80 (1996) 5593.
93. J. Kawanaka, A. Ogata, S. Kubodera, W. Sasaki and K. Kurosawa, Appl. Phys. B-Lasers and Optics 65 (1997) 609.
94. M. Kitamura, K. Mitsuka and H. Sato, Appl. Surf. Sci. 80 (1994) 507.
95. T. Nakamura, F. Kannari and M. Obara, Appl. Phys. Lett. 57 (1990) 2057.
96. A. EI-Habachi and K.H. Schoenbach, Appl. Phys. Lett. 72 (1998) 1.
97. U. Kogelschatz, B. Eliasson and W. Egli. J. Phys. IV France 7 (1997) C4-47.
98. Ch. K. Rhodes "Excimer Lasers", Vol. 30 of Topics in Applied Physics, Springer-Verlag, Berlin, 1984.
99. J.-Y. Zhang, H. Esrom, and I.W. Boyd, Appl. Surf. Sci. 109/110 (1997) 482.
100.J.-Y. Zhang, H. Esrom, and I.W. Boyd, Appl. Surf. Sci. 138-139 (1999) 315.
101.I.W. Boyd and J.-Y. Zhang,, Advanced Laser Technologies (ALT99), Potenza-Lecce, Italy, Sept 20-24, 1999.
102. T. Ohishi, S. Maekawa & A. Katoh, J. Non-Cryst. Solids 147&148 (1992) 493.
103.G.Q. Lo, D.L. Kwong, and S. Lee, Appl. Phys. Lett. 60 (1992) 3286.
104.C.H. An and K. Sugimoto, J. Electrochem Soc. 139 (1992) 1956.
105.J.-Y. Zhang, V. Dusastre, D.E. Williams and I.W. Boyd, J. Phys. D: Appl. Phys.32 (1999) L1.
106.N. Kaliwoh, J.-Y. Zhang and I.W. Boyd, Surface and Coating Technology 125 (2000) 424.
107.J.-Y. Zhang and I.W. Boyd, Jpn. J. Appl. Phys. 38 (1999) L393.
108.P. Singer, Semiconductor International, October 1994, p. 34.
109.S.P. Murarka, Solid State Technology 39 (1996) 83-90.
110.J.-Y. Zhang and I.W. Boyd, Optical Materials 9 (1998) 251.
111. C.A. Pryde, J. Polym. Sci.: Part A: Polym. Chem., 1989, 27, pp. 711-724
112. .J.-Y. Zhang and I.W. Boyd, E-MRS 99 Spring Meeting (to be published Appl. Surf. Sci. 2000).

**Lasers in Micromachining
and Surface Modification**

Mat. Res. Soc. Symp. Vol. 617 © 2000 Materials Research Society

PROCESS MODELING
OF THE LASER INDUCED SURFACE MODIFICATION OF CERAMIC SUBSTRATES FOR THERMAL AND ELECTRICAL LINES IN MICROSYSTEMS

Herbert Gruhn, Roland Heidinger, Magnus Rohde, Sabine Rüdiger, Johannes Schneider* and Karl-Heinz Zum Gahr*

Forschungszentrum Karlsruhe, Institute for Materials Research I,
*Universität Karlsruhe (TH), Institute of Materials Science and Engineering II,
P.O.B. 3640, 76021 Karlsruhe, Germany

ABSTRACT

Laser induced surface modification has been used to fabricate conducting paths in ceramic substrates. For the purpose of process simulation and prediction of process parameters a finite element model has been developed to simulate the thermal behaviour of the substrate during laser surface interaction. The results of the model calculation have been verified experimentally for alumina and Cordierite substrates. Using this model the width and the depth of the fabricated lines could be predicted as a function of the laser power and velocity. The stresses due to thermal mismatch are estimated and identified as the likely reason for crack formation which reduces the functionality of the conducting paths. Further developments will consider different ceramic substrates such as PZT.

INTRODUCTION

For many applications in microsystems technology the question of thermal management is of high interest. Very important are also robust conducting paths with a high electric conductivity. Today such paths between active elements on ceramic substrates are formed by lithographic methods in thin and thick film technology. A possible alternative is based on the laser induced surface modification [1] (see figure 1). Three different processes are considered. The remelting process is regarded as the fundamental process in which the substrate surface is melted locally by the laser beam and solidifies. In the injection process particles are directly sprayed into the laser melted surface. This is in contrast to the precoating process. In the first step the substrate is precoated with the particles to be dispersed. In a second step the laser beam remelts the substrate under the precoating and intermixes both. The cross-sections of solidified lines produced by the three processes in substrate material Cordierite ($2MgO \cdot 2SiO_2 \cdot 5Al_2O_3$) are presented in figure 1. To avoid cracking by thermal shock during laser treatment the ceramic substrates are preheated slightly below sintering temperature. The problem of the oxidation of metallic precoatings during substrate preheating was solved by the development of a special vacuum furnace which allows laser material treatment in vacuum.

The advantages of the processes are free design and a good bonding to the substrate particularly for higher temperature applications. The aim of this work is to fabricate small lines with a width down to 200 µm which show a significant increased thermal and electric conductivity compared to the ceramic. The ceramic substrate material considered in this study is Cordierite because of its outstanding properties such as low thermal expansion, good thermal shock resistance and low dielectric permittivity which is attractive for microwave circuits. Figure 2 shows a laser modified Cordierite substrate with a dispersed line of WC-particles produced with the precoating process. For this example line width d is 400 µm. A resistance R of 40 Ω was measured over a line length l of 8 mm. This is equivalent to a sheet resistance

$R_{sheet} = R \, b \, l^{-1} = 2 \, \Omega$ and corresponding to a resistivity $\rho(d)$ of $1.0*10^{-4} \, \Omega m$ ($\rho(d) = R_{sheet}*d$, line depth $d = 50 \, \mu m$). The resistivity of the ceramic substrate Cordierite is above $10^9 \, \Omega m$.

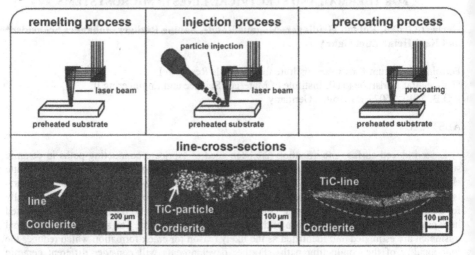

Figure 1. Laser induced surface modification

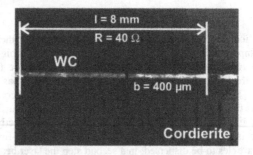

Figure 2. WC dispersed with the precoating process in substrate material Cordierite

MODEL DESCRIPTION

In order to support the selection and control of process parameters for laser induced surface melting a model was developed to consider the various aspects of the modification process. The model uses the finite element method (finite element program ABAQUS) and is based on the three-dimensional non-stationary non-linear equation of heat conduction. Radiation and convection are taken into account as well as temperature dependent material data for density, specific heat and thermal conductivity. Figure 3 illustrates which parameters influence the temperature distribution during the melting process.

Bath convection [2] which occurs during surface melting is responsible for the deviation from the semicircular cross-section shape. In dependence on the temperature derivative of surface tension of the melt ($d\sigma/dT$) two alternative cross-sections evolve. A negative derivative ($d\sigma/dT < 0$) means that the surface stress in the middle of the melt pool is much lower than at the borders. The melt is torn apart and in the centre particles flow from the bottom to the surface [3]. As a consequence a mass flow is formed in the bath as described in cross-section I (see figure 4). Thus the melt becomes elongated. Given a positive derivative ($d\sigma/dT > 0$) the mass flow follows the opposite sense, the melt extends deeper into the substrate and cross-section II is formed. Figure 4 also shows experimental cross-sections of alumina and Cordierite as examples for the two different pool shapes.

Since the bath convection leads to an anisotropic heat transport in the melt the FE-model uses anisotropic thermal conductivity data above the melting temperature of the substrate. To describe cross-section I the thermal conductivity is enhanced parallel to the surface with rising temperature. Shape II is approached by a highly reduced thermal conductivity parallel to the surface.

One of the main advantages of this model is that only two parameters have to be adjusted to predict cross-section width and depth during laser induced surface melting: The anisotropic part of the thermal conductivity above melting temperature and the amount of the CO_2-laser power absorbed by the substrate.

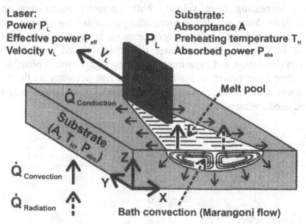

Figure 3. Thermodynamic model of laser induced surface melting

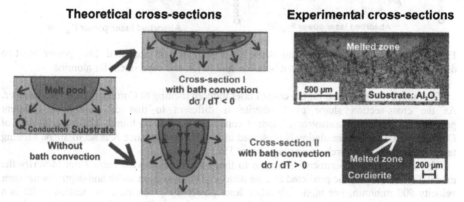

Figure 4. Dependence of the cross-section shape on bath convection

EXPERIMENTS AND SIMULATION

The model is first established for alumina for which the material data base as well as experimental experience is best founded. The temperature dependent materials data of alumina used for the simulation can be found in [4]. Melting temperature is 2323 K, preheating temperature 1773 K.

The free parameters of the model are determined by the comparison of width and depth of the cross-section between calculation and experiment for one single experiment [5]. The melt pool flow is approximated by the adjusted anisotropy of the thermal conductivity in x-direction above melting temperature. The absorptance A is set to 70 %.

For model verification it is necessary to compare simulation and experiment as a function of different experimental parameters. Therefore CO_2-laser power was varied between 45 W and 200 W for three different laser velocities: 250 mm/min, 500 mm/min and 1000 mm/min. Width and depth of the melted lines were measured. The experimental and calculated results are shown in figure 5. It is obvious that with increasing laser power cross-section width and depth are also increasing, with increasing laser velocity both geometry parameters are decreasing. As demonstrated in figure 5 the simulation successfully describes the experimental results in terms of cross-section width and depth. Their dependence on laser power and velocity is particularly well reproduced for higher power. If laser power is reduced towards minimal power the deviation between simulation and experiment is getting significant. A possible explanation could be that the effective laser power is determined with less accuracy as the CO_2-laser system is working at its lower limit. Alternatively the simple model is no longer fully adequate for lower laser powers and needs refinement.

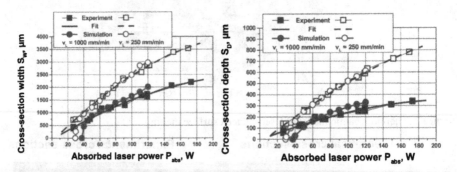

Figure 5. Cross-section width and depth in dependence on absorbed laser power for two different laser velocities: Comparison between simulation and experiment for alumina

The model was also used for the case of laser induced melting of Cordierite substrate surfaces. As the cross-section shape of Cordierite is different to that of alumina a different parameterisation of the anisotropic thermal conductivity must be found. The material data of Cordierite used for the simulation is presented in [4]. The absorptance A is set to 50 %. Melting temperature is 1738 K, preheating temperature 1573 K.

Also in this case experiments analogue to those for alumina were carried out to verify the model. Figure 6 shows the predicted and the measured cross-section width and depth for the laser velocity 500 mm/min. For higher absorbed laser power the predicted cross-section width is a good approximation of the experiments, for powers less than 40 W the calculated curve is clearly higher. This is equivalent to the results for alumina. Therefore the model adaptation must be further optimised. Further work is needed to verify the model by comparing the simulation to experiments made with other laser velocities.

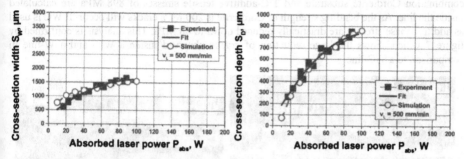

Figure 6. Cross-section width and depth in dependence on absorbed laser power: Comparison between simulation and experiment for Cordierite

The thermal and electrical conductivity and the bonding of the lines are significantly effected by cracks. Thermal shock and thermal mismatch are possible sources for crack formation.

If large heating and cooling rates occur in the laser melting process thermal shock becomes critical. Therefore the substrates are preheated to a high temperature to reduce the temperature differences during the laser process. For Cordierite as substrate material, in which the thermal shock resistance is especially high cause of its low thermal expansion, thermal mismatch is assessed to predominate.

The stresses by thermal mismatch appear during the cooling process from preheating to room temperature after the dispersed lines are produced. The different thermal expansions of substrate and line could evoke high stresses which are relaxed by crack formation. The stresses inside the dispersed line are estimated by equation 1 [5]. Table I specifies the corresponding material data.

$$\sigma_{Line} = \frac{E_{Line}}{1 - \nu_{Line}} (\alpha_{Substrate} - \alpha_{Line}) (T_{amb} - T_{plast}) \qquad (1)$$

Table I. Material data used for the simulation for different substrates and additives

Material parameter	Symbol	Unit	Cord-ierite	Al₂O₃	TiC	WC
Young's modulus	E	GPa	130	400	460	710
Thermal expansion coefficient	α	10^{-6} K⁻¹	3.6	8.0	7.95	3.9
Poisson number	ν	-	0.3	0.3	0.3	0.3
Bending strength	σ_b	MPa	200	400	-	-
Melting temperature	T_S	°C	1465	2050	3257	2785
Ambient temperature	T_{amb}	°C	20	20	20	20
Temperature above which stresses are reduced by plastic deformation	T_{plast}	°C	-	1000	-	-

For the estimation of the stress in the matrix material the values of E_{Line}, ν_{Line} and α_{Line} are assumed to be the mean of the values for the substrate and the dispersed material. For the combination Cordierite substrate and TiC-additive tensile stresses of 898 MPa are calculated inside the line. As the bending strength of Cordierite is 200 MPa cracks will occur. If WC is used as additive these stresses are drastically reduced down to 88 MPa. This analysis is confirmed by figure 7 which shows the large reduction of cracks observed inside a WC- and a W-line in comparison to a TiC-line.

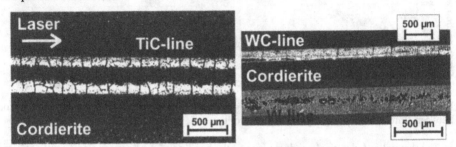

Figure 7. Top-view of a dispersed TiC-, a dispersed WC- and a dispersed W-line with cracks perpendicular to the laser moving direction, precoating process

CONCLUSION AND OUTLOOK

For the simulation of the laser induced surface modification a model was developed to calculate cross-section width and depth of the surface melting process. For the substrate materials alumina and Cordierite the results of the model calculation showed good agreement to experiments concerning the process parameters laser power and velocity. The model will be further used to find optimal process parameters to minimise the cross-section width. Still the model must be refined towards lower laser powers. Cracks inside the dispersed lines are mostly produced by thermal mismatch. The estimation of these stresses demonstrates that the stress amount is highly reduced by the use of WC as additive instead of TiC which is experimentally verified.

Further work is needed for the description of the dispersing particles (e.g. TiC, WC, W) introduced into the surface modification process. Experiments and simulations are being extended to PZT as an important material for actuator applications.

REFERENCES

1. K.-H. Zum Gahr, C. Bogdanow and J. Schneider, Friction and wear reduction of Al_2O_3 ceramics by laser-induced surface alloying, Wear, No. 181-183, pp. 118-128, 1995.
2. G. Tsotridis, H. Rother and E. D. Hondros, Marangoni Flow and the Shapes of Laser-melted Pools, Naturwissenschaften, No. 76, pp. 216-218, 1989.
3. G. Herziger, Lasertechnik, lecture reprint, Lehrstuhl für Lasertechnik, RWTH Aachen, Germany, 1994.
4. H. Gruhn, R. Heidinger, M. Rohde, S. Rüdiger, J. Schneider and K.-H. Zum Gahr, Proceedings Microsystems MSM 99, Puerto Rico USA, April 19-21,1999, pp. 105 - 108.
5. B. L. Mordike, Laseroberflächenumschmelzen von metallischen und nichtmetallischen Werkstoffen, Opto Elektronik Magazin, Vol. 4, No. 5, pp. 482 – 490, 1988.

Mat. Res. Soc. Symp. Vol. 617 © 2000 Materials Research Society

Laser micromachining of metallic mold inserts for replication techniques

W. Pfleging, A. Meier, T. Hanemann, H. Gruhn and K.-H. Zum Gahr
Forschungszentrum Karlsruhe GmbH, Institute for Materials Research,
P.O. Box 3640, 76021 Karlsruhe, Germany

ABSTRACT

The laser microcaving (LMC) of steel is performed with cw Nd:YAG laser radiation. LMC enables a "clean" patterning process with only a small amount of debris and melt. During LMC the formation of a Ni-enriched interface layer and an oxide surface layer may be observed. The formation of these reaction layers as well as the etch rate and the surface quality strongly depend on the chemical composition of the steel and the process parameters. Surface qualities with an roughness of about R_a(center line average)=300 nm can be realised. The ablation rates are in the range of 10^6 μm^3/s. With excimer laser radiation a further improvement of surface topographies can be achieved via laser planarisation. Mold inserts are manufactured by LMC, and microstructures composed of PMMA are successfully demolded by using the Ultraviolet light induced Reaction Injection Molding (UV-RIM) or Photomolding technique. CE(Capillary Electrophoresis)-Chips made of PMMA are successfully demolded, and the functionality of the CE-Chips is demonstrated.

INTRODUCTION

Microsystem technology has been the subject of research and development since the 1980s and will be a key technology for the new century. Different technologies have been developed, such as modified "classical" semiconductor technology using silicon [1], the LIGA technique [2], micromechanical abrasion or precise laser material processing [3,4]. The lateral dimensions of the generated microcomponents and systems are ranging from the nm to the mm scale depending on the method of manufacturing and on the type of application. For a necessary resolution of about several μm in lateral direction laser processing may be a suitable method for the manufacturing of microsystem components. But often laser beam processing is a slow serial step by step material removal by means of movement of laser beam and/or substrate. In order to realise a fast and reasonable large scale fabrication of microcomponents a laser material processing with respect to the manufacturing of mold inserts for micro injection molding techniques such as the Ultraviolet light induced Reactive Injection Molding (UV-RIM) designed for rapid prototype fabrication was developed [5-7]. Micro injection molding is an established and economic technology for the large scale fabrication of microcomponents made of polymers [8]. Main advantages of the laser material processing compared to the commonly used LIGA technique, spark erosion and micromechanical or chemical fabrication processes are a large choice of materials and a shorter processing time. The time consumption for manufacturing a mold insert via LIGA technique is generally in the range of several weeks up to several months depending on the structure height and structure complexity. The plating time of the nickel body alone is in the range of 2-3 weeks [9].

The use of laser radiation for the formation of molds made of polymer is extensively investigated by other groups [10,11,12]. They use excimer laser radiation either for the so called

Laser-LIGA or for the fabrication of mold inserts for metal injection molding. In the latter case the mold insert will be lost after demolding. Laser ablation with Q-switch Nd:YAG lasers also can be used for the 3d-microstructuring of cemented carbides. Applications as embossing die and punching tool were demonstrated [13,14]. In this work the rapid manufacturing of mold inserts for micro injection molding is realised by laser microcaving (LMC) of high alloyed and carbon steel. This technique is performed with Nd:YAG laser radiation and is a relatively new one [15] and is now used for the first time for the patterning of mold inserts. Mold inserts for micro injection molding must meet especially high demands with regard to surface roughness, because otherwise a demolding would be impossible. It will be shown that the material composition has a great influence on the achieved surface quality. We attend to offer a laser process technique to fabricate mold inserts within three days or less. The aim is a stable and reproducible process for the fabrication of molding tools. In this contribution results and applications of laser patterned metallic mold inserts are presented.

EXPERIMENTAL DETAILS

For LMC we use a solid state laser radiation source (SPECTRON Laser Systems SL300, Nd:YAG, wavelength λ=1064 nm) which operates in cw (continuous wave) mode. The laser beam is focused onto the sample surface by an objective lens (f=160 mm or 100 mm) and can be scanned over the sample surface via deflection mirrors up to speeds of 2000 mm/s. This enables us to treat an area of 110x110mm^2 (f=160 mm) or 65x65 mm^2 (f=100 mm). After LMC the metallic surfaces and structures are cleaned by using an excimer laser micromachining system (Exitech PS 2000). The laser radiation source (Lambda LPX 210i) operates at a wavelength of 248 nm. With Scanning Electron Microscopy (SEM), Energy Dispersive analysis by X-rays (EDX) and Auger Electron Spectroscopy (AES), the morphological structures and chemical compositions of the heat affected zones are investigated. The phase formation in the laser influenced areas is determined with X-ray deflection methods. Cross sections of laser ablated surfaces reveal the thickness of the reaction layers. With non-contact laser sensor technology (UBM MICROFOCUS) the etch rate and surface roughness is determined.

A simple description of the mechanisms during LMC is shown in figure 1: LMC is performed with laser powers of P_L=2-7 W in cw-mode. The laser intensity profile is a gaussian one with a focal diameter of 20 µm. The steel substrate is locally heated up by laser radiation. In order to pattern large areas the laser beam has to be scanned over the surface. In combination with oxygen as processing gas a laser induced oxidation of the steel surface occurs. Under special conditions the mechanical tensions inside the oxide layer reaches a critical value, and the oxide layer lifts off from the bulk material.

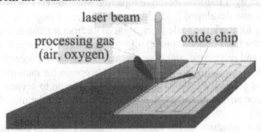

Figure 1. *Schematic presentation of laser microcaving (LMC) of steel materials*

RESULTS AND DISCUSSION

Reaction layer

Figure 2. *Cross sections of laser patterned surfaces*
(laser power P_L=4 W, laser scan speed v_L=5 mm/s)

The LMC process is much more complex than described by figure 1. For example the material composition of the steel substrate has a great influence on the formation and removal of the reaction layers. Figure 2 illustrates this fact: Three different kinds of steel materials show different formations of reaction layers while using the same laser and process parameters during LMC. Alloyed steel X4CrNi 1710 shows the formation of a dense oxide layer, which cannot be removed during patterning. X45NiCrMo4 reveals the formation of an oxide layer which is weakly bonded on the steel surface, and for carbon steel the oxide layers lift of from the bulk, and only a small heat effected zone remains on the surface. In case of alloyed steels containing Ni, an additional layer is formed between the oxide layer and the bulk material. Auger Electron Spectroscopy reveals, that the oxide layer is composed of Fe, O and Cr, while the interface layer reveals a significant Ni enrichment. The thickness of the interface layer is of about 1-5 µm and from EDX measurements we can conclude, that this layer contains about 70 wt% Ni. In the oxide layer $FeCr_2O_4$ and Fe_3O_4 phases dominate which can be concluded from X-ray deflection measurements. The reaction layers can easily be smoothed or removed by excimer laser material processing (figure 3). Small excimer laser fluences lead to a significant improvement of surface quality. With high excimer laser fluences the reaction layer is completely removed, and a smooth surface is remaining. In both cases, cracks in the surface will be completely removed. By this methods of planarisation an average surface roughness of about R_a=200 nm is attainable.

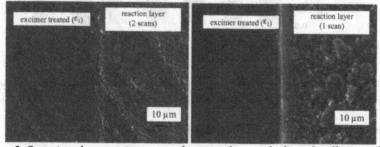

Figure 3. *Scanning electron microscopy of reaction layers which are locally treated with excimer laser radiation (material: X1CrNi2521; LMC: P_L=4 W, v_L=10 mm/s, one or two laser scans; excimer laser planarisation: ε_1=0.75 J/cm², ε_2=4.5 J/cm², 6 pulses)*

Figure 3 reveals, that the surface quality of the reaction layers induced via LMC is also a function of the number of laser scans. During the first laser scan (Figure 3, right) the laser beam is exposed to a clean steel surface, and the process of oxidation is initiated, while during the second laser scan (Figure 3, left) the laser beam is exposed to an oxide layer with changed properties concerning laser absorption. The oxide layer has an improved absorption of laser radiation compared to the clean steel surface, and a significant laser ablation during LMC occurs.

Ablation rate and surface quality

The following first results prove the capability of LMC for mold insert fabrication. The etch rate and the surface quality are mainly influenced by laser power P_L and scan speed v_L [16]. It is obvious that with increasing etch rate also the surface roughness increases. For a laser power $P_L=6.5$ W the etch rate increases up to 1.3×10^6 $\mu m^3/s$ for $v=20$ mm/s. Simultaneously the surface roughness increases up to $R_a=700$ nm. The smallest attainable surface roughness is $R_a=300$ nm without an additional post processing. Best results are obtained for scan speeds of $v_L=10$-20 mm/s and laser powers $P_L \leq 4$ W. Surface roughness and etch rate are also a function of material parameters as described in figure 4. Best results are realised with X8 CrNiMoNb 16-16. For this material high etch rates of about 1.2×10^5 $\mu m/s$ and a small surface roughness of about 300 nm is obtained. Finally, it has to be mentioned, that the process strategy during LMC is a main tool for the fabrication of metallic mold inserts. Process strategy stands for special technical measures like changing of laser scan sequence and direction, laser power variation, switching between Q-switch-mode and cw-mode and variation of laser incident angle. For example the process strategy decides if we obtain normal walls or slowly inclined walls during patterning of trenches. Steep walls are created by using small laser pulses with high pulse power while slowly inclined walls are formed by using low laser powers in cw-mode.

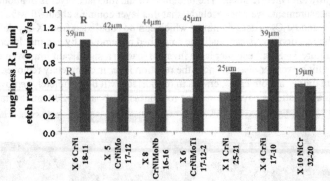

Figure 4. *Surface roughness R_a, etch rate R, and etch depth (values in diagram) for different steels at fixed process parameters (four laser scans, scan offset 5 μm; $P_L=3$ W, $v_L=10$ mm/s)*

Metallic mold inserts with CE-Chip designs

CE (Capillary Electrophoresis)-Chips are commercially made of glass, silicon or quartz [17]. They are used in biological and chemical analysis for the handling of small volumes of liquids (10^{-9}-10^{-12}l) in order to perform separation and detection (μ-TAS). At Forschungszentrum

Figure 5. *Metallic mold inserts with different CE-Chip designs fabricated with LMC (left: no post processing; center and right: post processing with excimer laser radiation)*

Karlsruhe GmbH and elsewhere [17] the use of polymer materials instead of Si, glass or quartz is investigated, because polymers can easily and economically be replicated in micro injection molding techniques. For these applications we attend to offer a laser processing technique which enables a fast and cheap fabrication of mold inserts compared to LIGA technique, spark erosion or micromechanical abrasion. With LMC mold inserts for CE-Chip applications with different designs and depth were fabricated (see figure 5). A detail view of an especially designed CE-Chip is shown in figure 6. In this case the channels have a cross-section of about 60 μm x 8 μm. The separation channels have a length of about 37.2 mm. The total area patterned during LMC is 38.8 mm x 14 mm. The CE-Chips are demolded with UV-RIM, and cross-sections up to 60 μm x 50 μm were realised. After demolding the channels have to be covered by an additional plate. The covering is performed with a special liquid adhesive. In order to avoid a contamination of the channels during covering, small adhesive grooves around the channels are necessary. With a special process strategy these grooves were realised in metallic mold inserts with a cross-section of about 1μm x 30 μm (see figure 6). Functional tests reveal that the CE-Chips are suitable for applications in analytical chemistry.

Figure 6. *Mold insert (left, post processing with excimer laser radiation) and demolded PMMA CE-Chip (right)*

CONCLUSIONS AND OUTLOOK

The combination of laser micromaching for the fabrication of metallic mold inserts and the subsequent replication via photomolding allows a rapid fabrication of microstructured parts made from polymer or composite materials. The total processing time starting from the CAD-data until the first prototypes can be reduced to three days or less. Further investigations will be focused on the following aspects:

- Further optimization of CE-Chip fabrication including the encapsulation of demolded polymer parts and the controlled fabrication of rough and smooth capillaries
- Characterisation of the demolded polymer modules concerning geometrical, physical and chemical properties necessary for applications in bio-analytics
- Further prototype molding using highly filled composite systems and fabrication of ceramic and metal prototypes
- Further refinement of laser process strategies adapted on three dimensional geometries.

ACKNOWLEDGMENTS

We are grateful to our colleague A. Sporrer for her technical assistance in SEM/EDX. We also thank M. Blumhofer for his support in laser profilometry and W. Bernauer and Dr. W. Hoffmann for helpful discussions. We gratefully acknowledge the financial support by the Deutsche Forschungsgemeinschaft (Mikromechanische Produktionstechnik, SPP1012).

REFERENCES

1. H. Jansen, H. Gardeniers, M. de Boer, M. Elwenspoek, J. Fluitman, *J. Micromech. Microeng.*, **6**, 14-28 (1996)
2. F. J. Pantenburg, S. Achenbach, M. Sesterhenn, *Wissenschaftliche Berichte des Forschungszentrum Karlsruhe GmbH*, FZKA **6080**, 77-82 (1998)
3. E. W. Kreutz, H. Frerichs, M. Mertin, D. A. Wesner, W. Pfleging, *Applied Surface Science*, **86**, 266-277 (1995)
4. W. Pfleging, A. Ludwig, E. Quandt, *Proc. 6th Int. Conf. on New Actuators*, 376-379 (1998)
5. W. Pfleging, V. Piotter, T. Hanemann, *Proc. European Conf. on Laser Treatment of Materials ECLAT 98*, 443-448 (1998)
6. T. Hanemann, J. H. Haußelt, R. Ruprecht, W. Pfleging, Werkstoffe für die Informationstechnik/Mikrosystemtechnik, (WILEY-VCH, 1999) pp.297-302
7. T. Hanemann, V. Piotter, R. Ruprecht, J. H. Haußelt, *American Chemical Society Symposium Series*, **706**, 67-75(1998)
8. V. Piotter, R. Ruprecht, J. Haußelt, *Microsystem Technologies* (1997) 129-133
9. W. Bacher, K. Bade, B. Matthis, M. Saumer, *Microsystem Technologies*, **4**, 117-119 (1998)
10. M. Abraham, J. Arnold, W. Ehrfeld, K. Hesch, H. Möbius, T. Paatzsch, C. Schulz, *SPIE Proc. Micromachining and Microfabrication Process Technology*, **2639**, 164-173 (1995)
11. R.A. Lawes, A.S. Holmes, F.N. Goodall, *Microsystem Technologies 3*, 17-19 (1996)
12. T. Shimizu, Y. Murakoshi, T. Sano, R. Maeda, *Microsystem Technologies*, **5**, 90-92 (1998)
13 A. Gillner, R. Poprawe, Werkstoffe für die Informationstechnik/Mikrosystemtechnik, WILEY-VCH Verlag GmbH, Weinheim, 359-364 (1999)
14. A. Gillner, *Proc. of the International Seminar on Precision Engineering and Micro Technology*, Aachen, Germany 105-112 (2000)
15. T. Abeln, J. Radke, F. Dausinger, H. Hügel, *Proc. European Conf. on Laser Treatment of Materials ECLAT 98*, 385-390 (1998)
16. W. Pfleging, T. Hanemann, A. Meier, *Surface Engineering Euromat* ,**11**, 455-460 (2000)
17. H. Becker, W. Dietz, *SPIE Proc. Microfluidic Devices and Systems*, 177-182 (1998)

Mat. Res. Soc. Symp. Vol. 617 © 2000 Materials Research Society

Quantification of Melt Ejection Phenomena during Laser Drilling

K.T.Voisey, C.F.Cheng & T.W.Clyne*
Department of Materials Science and Metallurgy, University of Cambridge, Pembroke Street,
Cambridge, CB2 3QZ, England
Tel: +44 (0)1223 334332 Fax: +44 (0)1223 334567
*Email: twc10@cam.ac.uk

ABSTRACT

During laser drilling, material removal in general occurs both by vaporisation and by the expulsion of molten material. The latter commonly arises as a result of the rapid build-up of gas pressure within the growing cavity as evaporation takes place, but the precise mechanisms responsible for the phenomenon are still unclear. The current work is aimed at gaining an insight into these mechanisms via measurements of the amount of material ejected from cavities during laser drilling under different conditions. Attention is first devoted to the issues which need to be considered when making experimental measurements of the fraction of material removed by melt ejection. These include the collection efficiency and the possibility of chemical changes occurring during the process. Results are then presented from work with a range of metallic substrates (mild steel, tungsten, copper, titanium, aluminium and nickel), drilled with a JK701 Nd-YAG laser under different conditions. Observed variations in the melt ejection levels have been studied for mild steel and aluminium and these are briefly considered in terms of the expected effects of certain material property values and the mechanisms of melt ejection. Results from an existing finite difference heat flow model are used to investigate the significance of melt ejection.

INTRODUCTION

Laser drilling is of considerable interest, since narrow, intense laser beams can be used to drill relatively deep, fine diameter holes, with little thermal or mechanical damage to the surrounding material. Industrial applications include drilling of cooling holes in turbine blades, to allow higher engine operating temperatures. There is also interest in drilling arrays of fine holes in aerofoil surfaces to reduce turbulence and associated drag[1]. Material removal can take place via two main routes during laser drilling; vaporisation and melt ejection. Melt ejection means removal of material in the condensed (usually molten) state during laser drilling. It is not well understood and the mechanisms involved appear to be complex. Since the (relatively large) latent heat of vaporisation does not need to be absorbed when material is removed by melt ejection, it could be regarded as an efficient drilling mechanism. In general, the energy required to remove material by melt ejection, even significantly superheated, is about one quarter of that which would be needed to vaporise the same volume. For example, to remove $1m^3$ of iron by vaporisation 65.8 GJ are needed whereas only 12.3 GJ are required for the removal of the same volume in the molten state[2]. On the other hand, melt ejection can result in irregular and poorly-controlled hole dimensions, particularly due to incomplete expulsion of melt[3], so that in this respect it may be undesirable.

Some reported observations concerning melt ejection have been obtained by high-speed photography[3-8]. This can be used to estimate the velocity of ejected particles, and possibly their size distribution, as well as giving information about the stage during the pulse period when ejection starts. Yilbas and Sami[8] suggested that their observations are consistent with melt ejection becoming initiated at about the point when nucleate boiling might be expected to start within the molten zone surrounding the hole. Luft et al[9] reported that a transition towards shorter, more intense pulses causes the nature of the melt movement to change from continuous liquid flow to atomisation, with inhibition of the recast layer near the hole entrance, and ultimately to suppression of ejection. There have also been suggestions[3] that melt ejection is limited at low power densities, leading to the concept that there is an intermediate range of pulse energies, for a given pulse duration, over which melt ejection is most pronounced. However, this is an area in which there is very little in the way of comprehensive and reliable experimental data.

Moreover, while several approaches to modelling of the transport phenomena involved in laser drilling have been outlined[10-16], the development of models incorporating reliable simulation of the melt ejection process is still in the embryonic stages. Von Allmen presented a 1-D steady state model[10], based on experimental observations of smooth walled holes. Melt ejection was included as the motion of a liquid under a pressure gradient. A threshold pressure gradient was required to overcome surface tension before melt ejection could occur. This model agrees quite well with experimental data concerning the effect of power density on the onset of melt ejection, but the predictions are incorrect at high power densities. Chan and Mazumder[11] treated laser drilling as a problem with two moving boundaries. Their results indicate a peak in the rate of material removal by melt ejection, as a function of power density. The model of Solana et al[16] incorporates two distinct mechanisms of melt ejection, one occurring early in the pulse period as a result of rapid pressure build-up and the second at a later stage by a more progressive process resulting from the pressure gradient associated with the radial decay in beam power. Some high speed photography observations were presented in support of this hypothesis.

Experimental data in the present study were obtained by collection of melt ejected material. Predictions from a previously-developed finite difference model[15], which does not incorporate the effect of melt ejection, are used to explore the conditions under which the phenomenon is most significant.

EXPERIMENTAL PROCEDURES

Laser Drilling Conditions

The laser used was a JK701 Nd:YAG, manufactured by GSI Lumonics (λ =1.06 μm). The configuration used for this work was based on a 120 mm focusing lens and the maximum power was 135 W. Beam quality measurements showed that the value of M^2 ranged between 22 and 38 for the laser pulses used here. Blind holes were drilled using single pulses of 0.5 ms pulse length, typical of Nd:YAG laser drilling. Pulse energies ranging from 0.5 to 2.6 J were used. The focused beam spot on the substrate surface was calculated, based on the beam quality measurements and geometrical optics, to be between 65 and 85 μm in radius, agreeing with experimental measurements of the laser damaged area. In the work reported here, holes were drilled in substrates of mild steel, aluminium, titanium, nickel, copper, zinc, tungsten and a nickel-based superalloy. The substrates were ultrasonically cleaned in acetone before drilling.

Measurement of Melt Ejection Fraction

To determine the proportion of the hole volume removed by melt ejection, a technique was developed to collect the ejected material. A thin glass slide was positioned in the beam path, above the surface to be drilled. As melt ejection occurred, the ejected droplets collided with and adhered to the slide. Since drilling often produces an area of debris around the entrance hole, the substrate was moved between each drilling. A similar requirement applied to the collection slide, since adhesion could be impaired by the presence of prior deposit material. A number of holes were drilled, using the same substrate and collection slide, both substrate and slide being moved, relative to the beam, between each drilling. The mass gained by the slide, due solely to material removed by melt ejection, and the mass lost by the substrate, due to all the material removed, were measured and the melt eject fraction (MEF) calculated as the ratio of these two values. While it is possible in principle for the measured weight gain to include a contribution from vaporised material condensed on the slide, in practice this effect is likely to be negligible - a conclusion supported by the appearance of typical deposits.

The slides used were ~ 0.1 mm thick, and weighed ~ 400 mg. Using a Power Wizard[TM] PW-250 conduction calorimeter, power loss on transmission through the slide was measured as about 5 %. In practice, the average power loss exceeded this, since it increased with progressive deposition of melt ejected material during drilling. However, further measurements indicated that it did not exceed 10%, even after drilling had been completed. To optimise droplet adhesion, the collection slides were dried for at least 40 hours in a desiccator prior to use. This minimised the thickness of the hydrated layer that always tends to be present on a glass surface, as well as ensuring that all slides were stored under similar conditions. It should, however, be recognised that the collection efficiency, estimated at 80 %, could vary between different substrate types.

Weighing Procedure

A Sartorius RC210P microbalance was used to weigh the samples and slides. This balance can nominally weigh to ± 10 µg although it is subject to thermal drift. The mass removed in the drilling of a single hole was of the order of 10–30 µg. To reduce the proportional errors in the calculated melt eject fraction, measured mass differences were increased by drilling between twenty and sixty holes in each substrate. This meant that the weight gain of a glass slide was of the order of 1 mg, i.e. a relative change of about 0.25%. It is estimated that this could be measured to a precision of around 10% on the value of the weight gain per drilling. The weights and weight losses of the substrates varied somewhat between substrate materials, but typical values were of the order of 1 g and 1 mg respectively. The error (associated with the weighing operation alone) in evaluating the weight loss per hole is also estimated at about 10%.

Effect of Droplet Oxidation

The mass gain of the collection slide was due to the droplets of molten substrate deposited on it. However further mass changes could occur due to oxidation (and possibly other chemical reactions) occurring in flight or after deposition. This was investigated using X-ray diffraction. Since the thickness of the deposited material was typically of the order of, or thinner than, the penetration depth of standard ($Cu_{K\alpha}$) X-rays (i.e. ~20-40 µm), X-ray spectra are representative of the deposits as a whole. Rietweld refinement[17] was used to determine the relative abundances of each phase present. This procedure involves comparison of predicted spectra with those obtained experimentally, refining a series of parameters, including the phase constitution, to optimise agreement. From the resulting composition, a correction factor was calculated for each material. Results from this operation, for the different metals examined, are shown in Table I. Multiplying the measured mass gain (Fig. 1) by this correction factor gives the actual mass gain of melt ejected material from which the MEF (Figs. 2 and 3) is then calculated.

Table I Phase proportions in melt ejected material, obtained by analysis of X-ray data.

Substrate	Phases present in deposit	Weight %	MEF correction factor
Aluminium	Al	100.0	1.00
Nickel	Ni	82.0	0.96
	NiO	18.0	
Titanium	Ti	5.8	0.76
	$TiO_{0.716}$	94.2	
Mild steel	Fe	35.0	0.84
	FeO	35.6	
	Fe_3O_4	29.4	
Tungsten	W	100.0	1.00
Copper	Cu	80.1	0.97
	CuO	19.9	
Zinc	Zn	100.0	1.00

RESULTS AND DISCUSSION

The dependence of the melt ejection fraction on pulse energy (for a pulse period of 0.5 ms) is shown in Fig.2 for mild steel substrates. Several features are apparent in this figure. Firstly, a significant proportion of the material removed has been displaced by ejection. This is in the approximate range of 30% to 60%. Secondly, while there is clearly some variability, there appears to be a trend for an initial increase in MEF as the pulse energy is raised (up to ~1.7 J), followed by a regime in which the proportion ejected decreases. This may be consistent with the concept of melt ejection being dependent on the rate at which the pressure within the hole builds up as vaporisation occurs and on the dimensions of the molten layer. If vaporisation does not happen sufficiently quickly, then the pressure is presumably inadequate to expel much liquid. If, on the other hand, the power density is very high, then the thermal gradients become very high, so that the molten zone is narrow, and this may lead to suppression of liquid motion.

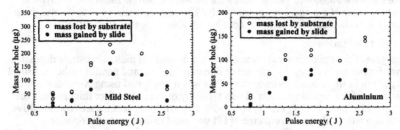

Fig.1 Measured mass changes of substrate and collection slide normalised by the number of holes drilled. Results are shown for aluminium and mild steel drilled with 0.5 ms pulses over a range of energies.

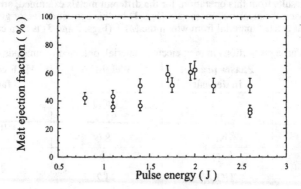

Fig.2 MEF as a function of pulse energy for steel substrates. The pulse period was 0.5 ms

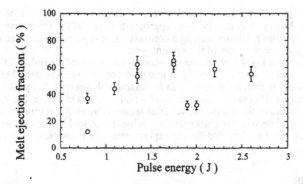

Fig.3 MEF as a function of pulse energy for aluminium substrates. The pulse period was 0.5 ms.

The above rationale might apply universally, although the ranges of pulse energy over which the transitions occur, and also the maximum and minimum levels of melt ejection, would be expected to depend on certain material properties and would thus vary for different metals. This was confirmed by the results obtained using other substrates. For example, Fig.3 shows that, while Al exhibits similar behaviour to mild steel, the MEF reaches peak values at rather lower beam energies. This may be partly associated with the relatively low density of Al, but its high thermal conductivity may delay onset of the ejection suppression regime, since it will favour a wide molten zone. It is clear that modelling of the melt ejection process itself, together with systematic experimental data for a wide range of substrates, are necessary in order to explore the relative importance of such effects. This is currently in hand, with the model being based on an existing finite difference formulation[15] for the heat flow and volatilisation characteristics.

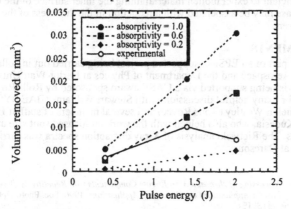

Fig.4 Volume removed per hole for mild steel, as a function of pulse energy, showing predictions from the numerical model (for three different absorptivity values) and experimental data obtained by weighing. The pulse period was 0.5 ms.

The existing finite difference model[15] has been used, in conjunction with the experimental data, to explore the variation of melt eject fraction with pulse energy (i.e. with beam power density). Fig.4 shows measured and predicted hole volumes as a function of pulse energy for mild steel. The predicted plots should represent lower bounds, since the model incorporates vaporisation but no melt ejection. Unfortunately, direct measurement of the surface absorptivity is quite difficult, particularly since it may change during the process. Its value may be quite low for a reflective metal during the initial period before the onset of surface melting. Subsequently it is likely to increase to a value close to unity as material excavation becomes established, although it may then be significantly depressed by the formation of an opaque plasma within the hole. It may thus be more appropriate to model the absorptivity as varying with time during the pulse, although using a constant value is probably justifiable for preliminary comparative purposes of the present type.

It looks likely from Fig.4 that an average absorptivity value of the order of 0.6-0.8 is appropriate for this case, which is broadly in line with expectation. The data then suggest that the contribution from melt ejection rises with increasing pulse energy, but perhaps starts to fall again at high energies. This is broadly consistent with the trends shown in Fig.2. Furthermore, the relative magnitudes of the additional volumes removed by melt ejection, indicated in Fig.4, are approximately in line with the corresponding values suggested by the data in Fig.2. This is encouraging, although it is clear that further experimental and theoretical work are needed before the effects of melt ejection can be reliably predicted.

CONCLUSIONS

The following conclusions can be drawn from this work:

1 In general, a significant proportion of the material removed during laser drilling of blind holes in metals is displaced by melt ejection. This proportion was found, using a precision weighing technique, to range up to about 70%, depending on the beam power density and the type of metal. Comparisons have been made between measured hole dimensions and predictions from a heat flow model which incorporates no provision for melt ejection.

2 For a given metal, a trend was noted for the melt ejection fraction to rise initially as the beam power density (pulse energy for a fixed pulse period) was raised, but then to fall. However, the details of the behaviour probably vary appreciably between different metals.

3 The initial increase in melt ejection on raising the beam power has been tentatively attributed to an increase in the pressure and pressure gradient generated within the hole by vaporisation, sufficient to expel molten material lining the inner surface of the hole. The subsequent decrease may be associated with a reduction in the thickness of the molten layer at high beam powers.

ACKNOWLEDGEMENTS

This work forms part of an EPSRC-supported project being carried out in collaboration with Rolls Royce, British Aerospace and the Department of Physics at Heriot-Watt University. One of the authors (KTV) is being supported via a CASE award sponsored by Rolls Royce. The authors are grateful for many helpful discussions with Stewart Williams (BAE SYSTEMS), Pamela Byrd & Jacquelyn Westley (Rolls Royce) and several members of staff at Heriot-Watt. These include Sean Kudesia, who also helped with the work on focused spot size and beam quality measurements. The Rietweld analysis of X-ray diffraction spectra was carried out using the EPSRC database at Daresbury.

REFERENCES

1. Bostanjoglo, G., et al. Processing of Ni-based Aero Engine Components with Repetitively Q-switched Nd:YAG Lasers. in High Power Lasers: Applications and Emerging Applications. 1996: Soc. Photo-Optical Instrumentation Engineers (SPIE).
2. Brandes, E. and G. Brook, eds. Smithells' Metals Reference Book. 7th Ed. ed. . 1992, Butterworth Heinemann: Oxford.
3. Korner, C., et al., Physical and Material Aspects in using Visible Laser Pulses of Nanosecond Duration for Ablation. Appl. Phys. A, 1996. 63: p. 123-131.
4. Murthy, J., et al. Investigation of the Drilling Dynamics in Ti-6Al-4V using High Speed Photography. in Laser Materials Processing Conference ICALEO '94. 1994. San Diego: Laser Institute of America.
5. Ramanathan, S. and M. Modest, High-Speed Photographic Studies of Laser Drilling of Ceramics and Ceramic Composites. J. of Laser Appls., 1995. 7: p. 75-82.
6. Yilbas, B.S., Study of Liquid and Vapor Ejection Processes During Laser Drilling of Metals. J. of Laser Applications, 1995. 7: p. 147-152.
7. Yilbas, B.S., A.Z. Sahin, and R. Davies, Laser Heating Mechanism Including Evaporation Process Initiating Laser Drilling. Int. J. of Machine Tools & Manufacture, 1995. 35(7): p. 1047-1062.

8. Yilbas, B.S. and M. Sami, *Liquid Ejection and Possible Nucleate Boiling Mechanisms in Relation to the Laser Drilling Process.* J. Phys. D: Appl. Phys, 1997. **30**: p. 1996-2005.
9. Luft, A., *et al.*, *A Study of Thermal and Mechanical Effects on Materials Induced by Pulsed Laser Drilling.* Appl. Phys A, 1996. **63**: p. 93-101.
10. Von Allmen, M., *Laser Drilling Velocity in Metals.* J. Appl. Phys., 1976. **47**: p. 5460-5463.
11. Chan, C. and J. Mazumder, *One-Dimensional Steady State Model for Damage by Vaporization and Liquid Expulsion Due to Laser-material Interaction.* J. Appl. Phys., 1987. **62**(11): p. 4579-86.
12. Ganesh, R.K., *et al.*, *A Model for Laser Hole Drilling in Metals.* J. Computat. Physics, 1996. **125**: p. 161-176.
13. Ganesh, R.K., A. Faghri, and Y. Hahn, *A Generalized Thermal Modeling for Laser Drilling Process: 1. Mathematical Modeling and Numerical Methodology.* Int. J. Heat Mass Transfer, 1997. **40**: p. 3351-3360.
14. Ganesh, R.K., A. Faghri, and Y. Hahn, *A Generalized Thermal Modeling for Laser Drilling Process 2. Numerical Simulation and Results.* Int. J. Heat Mass Transfer, 1997. **40**: p. 3361-3373.
15. Cheng, C.F., Y.C. Tsui, and T.W. Clyne, *Application of a 3-D Heat Flow Model to Treat Laser Drilling of Carbon Fibre Composites.* Acta Metall. et Mater., 1998. **46**: p. 4273-4285.
16. Solana, P., *et al.*, *Time Dependent Ablation and Liquid Ejection Processes during the Laser Drilling of Metals.* J. Phys. D : Appl. Phys., 2000: p. in press.
17. Young, R.A., *The Rietveld Method.* Monographs on Crystallography. 1993, Oxford: Oxford University Press.

Mat. Res. Soc. Symp. Vol. 617 © 2000 Materials Research Society

SURFACE PROCESSING AND MICROMACHINING OF POLYIMIDE DRIVEN BY A HIGH AVERAGE POWER INFRA-RED FREE ELECTRON LASER

MICHAEL J. KELLEY, Dept. of Applied Science, The College of William and Mary, Williamsburg VA 23187-7895, and Thomas Jefferson National Accelerator Facility, Applied Research Center, 12050 Jefferson Avenue, Newport News, VA 23606.

ABSTRACT

The long history and wide use of polyimide as a dielectric in the microelectronics industry has made it a favorite material for laser processing studies. The FEL used in the present work delivered picosecond-long 25 microjoule pulses at approximately 3.10 and 5.80 microns. The former is not associated with any strong absorption, while the latter is the strongest absorption band in the IR. This study explored hole drilling and surface transformation of as-made and aluminized DuPont Kapton* HN PMDA-ODA polyimde film.

INTRODUCTION

The opportunity for micromachining attracted the attention of the earliest investigators of laser ablation of polymers [1]. The need to cut small through-holes for vias in in printed wiring boards and for ink jet printer orifices focussed special attention on polyimide films. The research, development and deployment into production has been reviewed recently [2]. Polyimide's intense absorption throughout the UV [3] led to domination by excimer lasers. The discussion of the relative contributions of "true photochemistry" and thermal chemistry began at once [4]. The electronic structure of polyimides [5] is such that prospects are dim for exclusion of photochemical effects when UV irradiation is used. The success in manufacturing of ablative micromachining certainly questions the practical importance of the issue.

Pulsed laser deposition (PLD) also depends on ablation, making use as it does of the ejected material to form a film. Though inorganic materials continue to attract the vast majority of research interest, some attention has always been paid to polymers. Interestingly, reports on polymers having carbonyl groups (e.g, polyimides, polyamides, polyesters) are surprisingly sparse, despite their many important applications. It can be asked whether photochemistry leading to unwanted transformations (for PLD) is at least partly responsible. Previous results with polyamides [6] and polyimides [7] indicate that irradiation at strongly absorbed UV wavelengths results in elimination of the carbonyl leading to formation of polyamines in the first case and a carbonaceous network in the second. A potential ambiguity for polyimides is that purely thermal mechanisms, given sufficient time, can lead to a broadly similar result [8].

Polyimides thus are an appealing context in which to explore whether a suitable IR laser may both drive sufficiently rapid thermal processing to avoid pyrolysis on the one hand and avoid unwanted UV-driven photochemistry on the other. Absorption by polyimides in the infra-red is less intense than in the UV, but still significant. The strongest band of the familiar Kapton* H-series PMDA-ODA polyimide films appears at about 5.80 microns (1724 cm^{-1}) and is associated with the imide carbonyl, also the site of the UV activity. This has not been an easy wavelength to address with lasers that can deliver both the wavelength and high power. Free electron lasers (FEL's) offer a path; a recent review is available [9]. An FEL delivering more than 1 kW average power in the 3 – 6 micron range is now in operation at the Thomas Jefferson National Accelerator Facility [10].

EXPERIMENTAL WORK

We studied chiefly a commercial PMDA/ODA polyimide, DuPont Kapton* 100 HN as-made and an aluminized equivalent, Dunmore Type DE-320. Some materials were utilized as-received while others were rinsed with reagent-grade 2-propanol. No difference was found and we do not discuss it further. For laser irradiation, we attached small pieces of film at their edges to rigid metal plates and mounted these in a micrometer-driven positioning fixture.

The FEL was operated off peak tuning to deliver a FWHM peak width of less than 0.05 microns (< 0.9%) at 5.80 and 3.10 microns, compatible with the absorption peak width of the material. The laser output comprised a 18.75 MHz stream of 1 ps micropulses, delivered chiefly as 10 microsecond bursts ("macropulses") at 2 Hz for the work discussed here. CW operation, other macropulse lengths and other macropulse pulse repetition frequencies were also explored. The 2.4 cm nominally Gaussian beam was apertured to 2.2 cm and then focussed by CaF_2 lenses (Janos), having a focal length of either 150 mm or 62 mm. Actual delivered powers were measured with a Molectron power meter after all optics by increasing the macropulse repetition rate to 60 Hz to obtain sufficient signal for accuracy.

We measured the energy per macropulse as 430 mJ, or 2.29 mJ per micropulse if all were equal. In fact, however, the FEL is turning on during the first part of the macropulse, so that the initial micropulses are smaller and the latter ones larger. However, this cannot cause any macropulse to deviate more than 50% from the nominal value above. The beam waists at 3 μm are calculated to be 60 and 25 microns for the 150 mm and 62 mm lenses respectively. Taking the micropulse duration as 1 ps leads to estimated peak powers of 8.2×10^{10} and 1.9×10^{11} W/cm^2 respectively. Note that no account is here taken of the nominally Gaussian beam profile. These energies were chosen to bring the power per unit volume of irradiated material into the range of the excimer laser experiments, where the absorbances are about two orders of magnitude higher.

After the optical conditions were set, we delivered the chosen pulse sequence without moving the specimen: percussion drilling. For each set of optical conditions, wavelength and power, we ran a through-focal series with 250 micron axial increments at successive locations. We enclosed the complete experiment in a nitrogen-purged glovebag for 5.80 micron operations. To gain further insight into the ejected material, we placed a multi-layered roll of the film at the target position of a PLD chamber, about 10 mm distant from a quartz plate. Operating at 5.80 microns, we attempted to traverse the beam rapidly over the film, but noticed upon removal that the irradiated area was blackened.

We first examined all materials optically at 60 X and then later by SEM. To get the most sensitive insight into the effect on surface chemistry, we examined selected materials by Time-of-Flight Secondary Ion Mass Spectrometry (ToF/SIMS) using our Physical Electronics TRIFT-II CE. Though perhaps less familiar than some methods of surface analysis, there is available both a thorough review of the technique [11] and of its applications to polymer surfaces [12]. We operated the gallium ion gun at 15 kV, rastered over a square between 200 and 400 microns on a side, as needed to avoid charging. The instrument software provided for retrospectively selecting data from the near-hole region.

RESULTS

Much more was found than can be presented in detail in this brief report. The optical microscopy showed that at 5.80 microns, for either lens and for micropulse energy at the specimen between 13 and 2 µJ/micropulse, a single 10 µs macropulse always produced a through-hole at focus with little or no evident change in material adjacent to the hole. The higher the micropulse energy, the further from focus we could make single pulse through-holes. The next focal step beyond complete penetration sometimes showed pits, but often showed nothing. No blackened material was ever evident. The principal effect of increasing from a single macropulse per hole to 5 or 10 macropulses was to produce a ring of visually darkened material or a spot of the same where the single macropulse failed to produce a through-hole. Prolonged exposures, well in excess of 100 macropulses, resulted in a large blackened area. However, the relatively thin metal plate behind the film was also heated under these conditions, perhaps then heating the adjacent film. Changing wavelength to 3.10 microns resulted in incomplete penetration under most conditions. Material remaining in the hole incandesced with successive macropulses and eventually disappeared. Nothing was evident optically on the substrate used in the PLD experiment.

Figs.1-3 present ToF/SIMS results. Kapton* Type E film has previously been characterized by others using ToF/SIMS [13] as part of research into the effects of plasma processing. The principal difference is the inclusion of small amounts of other co-monomers in Type E, but the main conclusions should be pertinent. In particular, they found that the positive ion spectra in all conditions were not distinctive among themselves or compared to other polymers. However, the negative ion spectra showed fragments at masses 118, 134 and 144 that were highly characteristic of polyimide.

Fig.1 shows the similarity of the positive ion spectra from the untreated regions of two different samples and the effect of 10 macropulses at focus at 5.80 or 3.10 microns. The only substantial difference is at mass 73, but this is also a characteristic fragment of polydimethyl siloxane, a nearly ubiquitous silicone contaminant. Fig.3 shows negative ion spectra from the same samples. A slightly shifted version of the peaks reported for the Type E film is evident. For one sample, significant additional peaks are evident at mass 127 and 129, which are not removed by irradiation. They were not seen in the work with Type E film. In the ToF/SIMS apparatus, some areas of the PLD substrate were covered with a deposit, while others showed nothing. Fig. 2 shows negative ion spectra comparing treated and irradiated material to the material on the PLD substrate. The deposits show strong peaks at mass 127 and 129, and traces of the other masses associated with the polyimide. Material deliberately over-irradiated to appear so as to be visually black showed none of these peaks.

CONCLUSION

Though the present results must be viewed as highly preliminary, it seems evident that an IR laser operating at an absorption resonance has potential for development as a micromachining tool. As expected, no evidence is seen for photochemisty, in contrast with UV lasers. Further, no evidence of thermal transformation was seen either under the most favorable conditions for micromachining. IR-driven laser ablation at strongly-absorbed wavelengths may offer unique opportunities that deserve further investigation.

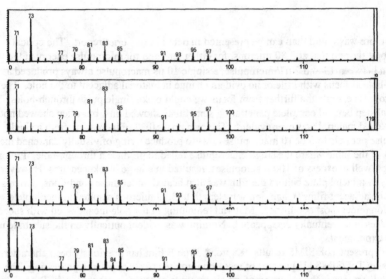

Figure 1)   Kapton HN, positive SIMS. a) Sample K11 untreated area. b) Sample K4 untreated
area. c) Sample K4, 5.80 microns. d) Sample K11, 3.20 microns.

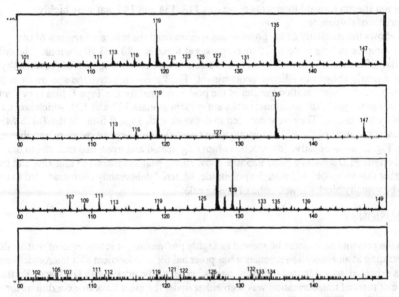

Figure 2): Kapton HN sample K4, negative SIMS a) untreated area. b) 5.80 micron. c) PLD
deposit material . d) Adjacent deposit-free PLD substrate.

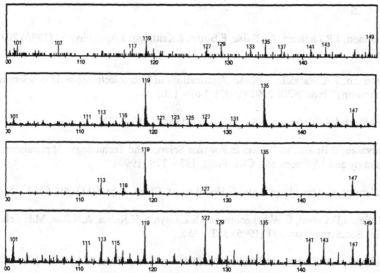

Figure 3) Kapton HN, negative SIMS. a) Sample K11 untreated area. b) Sample K4 untreated
area. c) Sample K4, 5.80 micron. d) Sample K11, 3.20 micron

ACKNOWLEDGEMENT

A great debt is owed for the help of others in this work: Michelle Shinn and George Neil of
Jefferson Lab with the FEL operation and beam characterization, Prof. Anne Reilly and Jason
Gammon of William and Mary for the use of the PLD apparatus and help with its operation, and
Amy Wilkerson of William & Mary for operation of the ToF/SIMS.

This work was supported by the U.S.Department of Energy by contract DE-AC05-84ER40150
under which the Southeastern Universities Research Association (SURA) operates the Thomas
Jefferson National Accelerator Facility. Additional support was provided by the Office of Naval
Research, the Commonwealth of Virginia and the Jefferson Lab Laser Processing Consortium.

REFERENCES

1. J.H.Brannon, J.R.Lankard, A.I.Baise, F.Burns, J.Kaufman; J.Appl.Phys.58 (1985) 2036 – 2043.

2. J.H.Brannon, T.A.Wassick; in "Laser Applications in Microelectronic and Optoelectronic Manufacturing" Proc. SPIE 2991 (1997) 146 – 150.

3. E.Sutcliffe and R.Srinivasan; J.Appl.Phys.60 (1986) 3315 – 3322.

4. R.Srinivasan, B.Braren; in "Lasers in Polymer Science and Technology: Applications", P.Fogarassy and J.F.Rabek eds, CRC Press, 133 – 179 (1990).

5. J.P.LaFemina, G.Arjavalingam and G.Hougham; J. Chem. Phys. 90 (1989) 5154 – 5160.

6. M.J.Kelley, J.D.Cohen, C.W.Erkenbrecher, S.L.Haynie, H.Kobsa, A.N.Roe, M.H.Scholla; Mat.Res.Soc.Symp.Proc. 397 (1996) 353 – 356.

7. J.L.Hohman, K.B.Keating, M.J.Kelley; Mat.Res.Soc.Symp. Proc. 354 (1995) 571 – 577.

8. P.V.Nagarankar, E.K.Sichel; J.Electrochem.Soc.136 (1989) 2979 – 2982.

9. H.P.Freund, G.R.Neil; Proc.IEEE 87 (1999) 782 – 803.

10. S.V.Benson; Proc.1999 IEEE Part. Accel. Conf. 1 (1999) 212 – 216.

11. A. Beninghoven; Angew.Chem.Int. Ed.Engl. 33 (1994) 1023 – 1043.

12. K.Wien; Nucl.Instr.Meth.Phys.Res.B 131 (1997) 38 – 54.

13. D.Wolany, T.Fladung, L.Duda, J.W.Lee, T.Gantenfort, A.Benninghoven; Surf.Interface Anal. 27 (1999) 609-617.

**Laser-Based Deposition
of Oxides**

Mat. Res. Soc. Symp. Vol. 617 © 2000 Materials Research Society

Growth of ZnO/MgZnO Superlattice on Sapphire

J.F. Muth, C.W. Teng, A.K. Sharma[1], A. Kvit[1], R.M. Kolbas, J. Narayan[1]
Department of Electrical and Computer Engineering, North Carolina State University,
Raleigh, NC 27695
[1]Department of Materials Science and Engineering, North Carolina State University,
Raleigh, NC 27695-7916.

ABSTRACT

The optical and structural properties of ZnO/ MgZnO superlattices were
investigated by transmission electron microscope, transmission measurement and
photoluminescence. The uncoupled wells ranged in thickness from ~30 Å to 75 Å.
Modulation of the Mg content was observed in Z-contrast TEM indicating the alloy
composition was periodic. The density of stacking faults in the superlattice was
extremely high, however the photoluminescence in the narrowest well case was blue
shifted, and substantially brighter than comparable bulk layers of ZnO and MgZnO
indicating that the emission was enhanced. Excitonic features were observed in the
optical absorption spectra and also revealed that diffusion of Mg from the barrier layers
into the well was occurring.

INTRODUCTION

A great deal of research has been performed in wide band gap semiconductors
resulting in the commercialization of group III nitride blue lasers, blue and green light
emitting diodes and ultraviolet photodetectors for use in display optical data storage and
solar-blind detection applications.[1] As an alternative to the GaN material system ZnO
and its alloys are of substantial interest. There are many similarities between GaN and
ZnO, they are both wurtzite and have similar band gaps, both exhibit strong excitonic
emission. However the exciton binding energy is nearly 3 times as large in ZnO (~60
meV) which permits makes excitonic effects even more pronounced.[2] As yet, p-type
doping of ZnO not technologically feasible although some reports indicate that nitrogen
may act as an acceptor.[3]

We have recently been focusing on the growth of MgZnO alloys to investigate
the potential of bandgap engineering the ZnO material system. While the equilibrium
solubility of Mg in ZnO is ~2 percent through pulsed laser deposition we have been able
to achieve metastable alloys with Mg concentrations of up to 36 percent.[4] The
absorption and photoluminesence spectra indicated that the exciton persists despite
alloy broadening at room temperature. These alloys have been shown to be thermally
stable for temperatures less than 700 °C indicating that formation of stable
heterojunction interfaces should be practical.[5] A super lattice structure comprised of
ZnO and $Mg_{0.2}Zn_{0.8}O$ has also been demonstrated by Ohtomo et al., indicating that ZnO
alloy based quantum structures should be feasible.[6]

In this work we report on the growth of a MgZnO superlattice by PLD. The superlattice was characterized by high-resolution transmission electron microscopy, transmission measurements and photoluminescence. In optical transmission measurements, the excitonic features of the absorption were enhanced and slightly blue shifted. The photoluminescence from the sample was very bright and blue shifted from the corresponding ZnO band edge value. While several samples were examined optically, only one sample has been analyzed by TEM. In transmission electron microscopy, the z-contrast technique indicated that the Mg content was modulated according to the expected period of the super lattice. High-resolution transmission microscopy revealed numerous horizontal stacking faults, and that the interface between the MgZnO barriers and ZnO wells was poorly defined. The well thickness was also wider than expected lessening the confinement, which complicates the analysis. However, we found the study illuminating since it provides insight into the broadening mechanisms and growth issues that are expected to be important in the growth of quantum wells and heterostructures in this material system.

EXPERIMENTAL DETAILS

The MgZnO superlattice and bulk films in this study were deposited by pulsed laser deposition on c-plane double-side polished sapphire. Before deposition the sapphire was cleaned in an ultrasonic bath using acetone and methanol. The vacuum system was evacuated to ~5x 10^{-8} Torr and the substrate temperature was maintained at 650 °C during deposition. The low temperature was intended to minimize Mg diffusion. During growth a pulsed KrF excimer laser (λ=248 nm, pulse width=25 ns, and repetition rate=10 Hz) with laser energy densities in the range of 3-4 J/cm^2 was used to ablate MgZnO and ZnO sintered targets. The composition of the MgZnO target was the same as that used to grow $Mg_{36}Z_{64}O$ bulk layers. A 1000 Å buffer layer of ZnO was deposited first to promote a smooth growth mode. Then 10 alternating layers of ZnO, ranging from ~30-75 Å in thickness and MgZnO (~120 Å) were deposited. The thickness of the barrier layers was chosen to ensure that the wells would be uncoupled. The thickness of the well was controlled by deposition time, with the rate determined from the growth of thicker films.

RESULTS AND DISCUSSION

TEM of the ZnO/sapphire and MgZnO/Sapphire interfaces showed that the first 100 Å of ZnO or MgZnO had a very high density of defects. High resolution TEM also indicated the formation of a spinel phase and the interface boundary in the case of the ZnO/Sapphire interface. After the first 100 Å the defect density was greatly reduced and improved as the film thickness was increased. Threading dislocations were the dominant defect in MgZnO bulk layers as shown in Figure 1. Threading dislocations were observed to propagate from the interface to the surface. Some stacking faults were also observed in the bulk MgZnO layer.

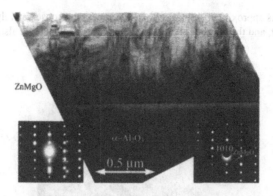

Figure 1. TEM of bulk MgZnO alloy. Threading dislocations dominate, while some stacking faults are visible in the upper left corner. The insets show the epitaxial relationship between the sapphire and MgZnO layer is maintained.

The superlattice structure is dominated by stacking faults as shown in Figure 2. Additionally, a higher density of stacking faults is found in the ZnO layer. The density of stacking faults made it very difficult to detect interfaces between ZnO wells and MgZnO barriers. However using Z-contrast technique it was found that the brightness was modulated periodically with the periodicity expected, indicating that the Mg concentration was not interdiffusing to the point of giving a homogeneous alloy in the superlattice region. The interfaces of the one sample examined were certainly not clearly defined, which make independent confirmation of the well thickness by TEM impractical.

The absorption spectra for three super lattices, ZnO, and $Mg_{36}Zn_{74}O$ at room temperature are shown in Figure 3. The relative absorbance is scaled such that the "A" exciton peak is the same for each sample. In the case of the 34 % Mg sample the thickness of the sample was such that the above band gap absorption was approximately the same as the other samples. In the ZnO sample the "A" and "B" excitons are very apparent, with the "A" exciton being very pronounced. As the deposition time of the ZnO well is decreased from 18 to 10 seconds (~80 to 30 Å in well thickness) the barrier layer composition decreases from ~29% to ~19 percent. The line width of the exciton resonance in the barrier layer also decreased as the well width was increased, indicating that alloy broadening was increasing with the width of the wells.

The "A" exciton resonance of the ZnO well layers appears to be more sensitive to the broadening mechanisms than the "B" exciton. With the increased well widths the "B" exciton became much more pronounced at room temperature. The absorption spectra consistently shifts to higher energies as the well width is decreased. The formation of low percentage Magnesium, MgZnO alloys would also be expected to blue shift the absorption spectra. Relatively little is know about this material system and while the amount of strain and magnitude of piezoelectric effects are expected to be less than that of the GaN/AlGaN system they should not be discounted. The blue shift of the narrowest well width is approximately 3 times that of what can be conservatively estimated for the strain effect, and the number of defects present in the film should also relax the thin film.

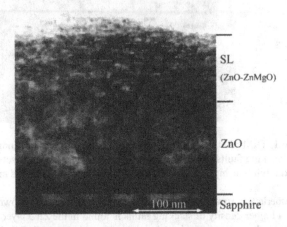

Figure 2. TEM image showing large number of stacking faults in the superlattice region. A higher than normal number of stacking faults is also apparent in the ZnO buffer layer.

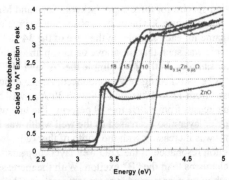

Figure 3. Absorbance of superlattice films, bulk MgZnO and bulk ZnO film scaled to "A" exciton absorbance. Note that the barrier layer composition decrease as the well deposition time increases. The 18 sec well is very similar to bulk ZnO, However the 15 and 10 sec wells are blue shifted.

The photoluminescence (PL) spectra of ZnO and ZnO/MgZnO superlattice films, at room temperature, excited by the 270 line of an Ar$^+$ ion laser is shown in Figure 4. The luminescence of bulk ZnO is usually very bright. The relative intensity of the PL of the superlattice films was substantially brighter, but was not quantitatively measured. The PL of bulk MgZnO and ZnO films is also usually red shifted with respect to the optical absorption edge. In the 10 and 15 second well thickness films the PL was blue shifted. In the PL of the 7 sec well sample, the barrier layer luminescence, and the ZnO emission from the buffer layer are observed. In this sample, we believe no enhancement of emission was obtained. In the 15 sec sample, the buffer layer is visible as a shoulder on the long wavelength side, enhance emission at about 362 nm and the barrier layer PL is apparent at ~340 nm. The 10 sec sample, with emission at ~360 nm, has a line width comparable to that of the bulk ZnO film, and the buffer layer emission was not apparent. To explain this PL, we conjecture that while clearly defined interfaces are not visible in TEM, and numerous stacking faults are present, confinement effects probably enhance the emission. A counter argument is that the emission is from lower concentrations of MgZnO formed by diffusion of the Mg from the barriers into the well regions. The TEM indicates that the alloy formed in this manner is certainly not homogeneous. The Z contrast TEM indicates that spatial variation is on the order of sizes where quantum effects should start to have an influence. In this case the wells are not simple square wells, but confinement effects could still provide enhancement.

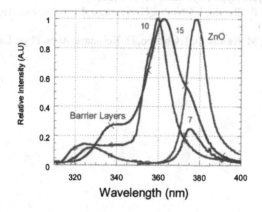

Figure 4. Intensity of PL spectra for MgZnO superlattices. The 10sec, and 15 sec wells, and bulk ZnO are scaled to 1 for clarity. The actually efficiency of the 10 and 15 sec superlattices was actually significantly brighter than that of the bulk ZnO. The PL of the 7 sec wells was diminished in comparison and is shown proportionally.

CONCLUSION

In conclusion, the optical and structural properties of ZnO/ MgZnO superlattices were investigated by transmission electron microscope, transmission measurement and photoluminescence. The uncoupled wells ranged in thickness from ~30 Å to 75 Å. Modulation of the Mg content was observed in Z-contrast TEM indicating the alloy composition was periodic. The density of stacking faults in the superlattice was extremely high, however the photoluminescence in the narrowest well case was blue shifted, and substantially brighter than comparable bulk layers of ZnO and MgZnO indicating that the emission was enhanced. Excitonic features were observed in the optical absorption spectra and also revealed that diffusion of Mg from the barrier layers into the well was occurring.

REFERENCES

1. S.J. Pearton, J.C. Zolper, R.J. Shul, and F.Ren, J. Appl. Phys **86**, 1 (1999)
2. J.F. Muth, R.M. Kolbas, A.K. Sharms, S. Oktyabrsky, and J. Narayan J. Appl. Phys. **85**, 7884, (1999)
3. J. Mathew, H. Tabata, T. Kawai, Jpn. J. App. Phys. Part 2, **38**, L1205 (1999)
4. C.W. Teng, J.F. Muth, Ü. Ögür, M.J. Bergmann, H.O. Everitt, A.K. Sharma, C.Jin, and J. Narayan, Appl. Phys. Lett. **76**, 979 (2000)
5. A. Ohtomo, R. Shiroki, I. Ohkubo, H. Koinuma, M. Kawasaki, Appl. Phys. Lett. **75**, 4088 (1999)
6. A. Ohtomo, M. Kawasaki, I. Ohkubo, H. Koinuma, Appl. Phys. Lett. **75**, 980 (1999)

Pulsed-Laser Deposition

Mat. Res. Soc. Symp. Vol. 617 © 2000 Materials Research Society

Optimization of the processing parameters for pulsed laser deposition of nickel silicide ohmic contacts on SiC

C. J. K. Richardson, M. H. Wisnioski, J. B. Spicer
Department of Materials Science & Engineering, The Johns Hopkins University, Baltimore, MD 21218
J. D. Demaree, M. W. Cole, C. W. Hubbard, P. C. Joshi, J. K. Hirvonen
Weapons and Materials Research Directorate, Army Research Laboratory, APG, MD 21005
H. Kim, A. Pique and D. B. Chrisey
Naval Research Laboratory Code 6370, Washington DC 20375

ABSTRACT

This research investigates the potential of pulsed laser deposition to create reliable high current ohmic contacts of Ni_2Si on single crystal 4H-SiC. Since this stoichiometry is the stable interphase in the nickel-silicon carbide diffusion couple, direct deposition eliminates the detrimental excess carbon normally formed by direct sintering Ni on SiC, the surface roughening that results from this sintering as well as the need for post-deposition high-temperature (900°C) anneals that are required in complex multi-component contacts. This study examines the processing parameters that must be used during deposition to obtain the desired microstructural characteristics for the contact. Pulsed laser deposition of nickel silicide produces smooth films with an amorphous or nanocrystalline structure interspersed with macroparticles. Macroparticle formation on the resulting films appear in the form of solidified droplets of the eutectic composition nickel silicide (3:1) that form during the long term target processing. The dependence of the number and size distributions of these droplets on laser fluence sample temperature is examined.

INTRODUCTION

High quality single crystal silicon carbide (SiC) has enormous promise as a large bandgap semiconductor for applications in high-temperature, high-current load applications.[1] The promise of SiC for these applications arises from its material properties and its ability to be processed by the same techniques as silicon. Perhaps the largest obstacle impeding the practical application of SiC is the inability to fabricate stable ohmic contacts. The difficulty in fabricating stable metallizations on SiC for pulsed power applications evolves from the electrical requirements of matching band structures, while maintaining a physically distinct contact. Most metals form silicide or carbides, thus making it very difficult to maintain device structure integrity at elevated temperatures.[2] Previous research has revealed nickel-based metallizations as having the most potential. Unfortunately, nickel reacts with SiC to produce Ni_2Si[3] which leaves excess carbon through the interaction region.[4] During this reaction the interface becomes rough and in some cases voids may form.[5]

By directly depositing Ni_2Si, the thermodynamically stable interphase between Ni and SiC, onto 4H-nSiC an ohmic contact can be produced without producing the unwanted free carbon. To achieve this goal, Ni_2Si was deposited on SiC using pulsed laser deposition (PLD).

PLD is an extremely versatile deposition technique that offers high levels of compositional control along with low-temperature deposition capability. Nickel silicide films were deposited onto single crystal 4H-nSiC in two different deposition chambers in order to study the effect of laser fluence, and substrate temperature. In particular, the structure, composition and interface boundary morphology of films from three separate depositions have been analyzed with scanning electron microscopy (SEM), electron diffraction spectroscopy (EDS), x-ray diffraction (XRD) and Rutherford backscattering analysis (RBS). The electrical behavior of these films has been determined by measuring the through thickness electrical resistivity character of the Ni_2Si – SiC interface using a two-point probe method.

EXPERIMENTAL DETAILS

Ohmic contacts of high purity Ni_2Si (99.9%) were created on research grade, s-face, 8° off-c-axis, select micropipe density 4H-nSiC (from Cree Research Inc) with a resistance of 0.020 ohm-cm. Ni_2Si was deposited under a base vacuum pressure of less than 1×10^{-6} torr using either a Lambda Physik LPX-300 or Lambda Physik Compex 205 excimer laser. When using the LPX-300 the laser beam was directed through a pinhole filter in order to remove hot spots. The focused beam was continuously rastered across a rotating Ni_2Si target with a 45° incident angle; the sample-target separation was 7 cm and the laser fluence was 2.3 Jcm^{-2} with a repetition rate of 10Hz. When using the Compex 205 laser, the beam was directly focused onto the rotating Ni_2Si target at a 45° incident angle. In this case the sample-target separation was 10 cm and the laser fluence was 10 Jcm^{-2} and a repetition rate of 50 Hz. Films that are a few hundred nanometers thick were grown on grafoil, (100) Si and 4H-nSiC at room temperature. Nickel silicide films were also grown on Si and SiC at 500°C. In both cases Si and grafoil were used as inexpensive test substrates in order to understand the deposition effects on the properties of the nickel silicide film.

Analysis of the nickel silicide films was done on films deposited on all three substrates. The depth dependent composition was determined through Rutherford backscatter spectroscopy analysis that was conducted with a NEC Pelletron accelerator, using a 2MeV He+ ion beam and a backscattering angle of 170°. An ETEK scanning electron microscope was used for morphology and EDS analysis. A Siemens D5005 x-ray diffractometer was used to determine the microstructure of the nickel silicide films. The two point probe electrical measurements were completed using a HP 4140B semiconductor test system with the use of a Ga-In eutectic to produce a low resistivity ohmic contact on the backside of the SiC wafer.

DISCUSSION

Heating the SiC substrate during deposition causes the most dramatic difference between the various films. For the samples that were not heated, the films were amorphous or had a nanocrystalline nature. This is evident in **Figure 1** where the width of the main diffraction peak is significantly broadened. **Figure 1** compares the XRD data from sample A which was not heated during deposition with the same angular scan of sample C, which was maintained at 500°C during deposition. It is clear from the narrow width of the diffraction peak and the emergence of the diffraction peaks near 43.6° and 48.9° that sample C is polycrystalline

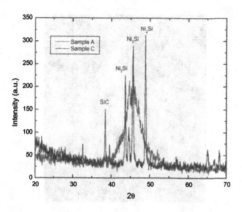

Figure 1 X-ray diffraction comparison of Sample A (gray) which was deposited at room temperature and Sample C (black) which was deposited at 500°C.

That an amorphous Ni$_2$Si film resulted from PLD onto a room temperature substrate is not surprising considering that it is known that silicon typically is deposited as an amorphous film. From a device perspective, the presence of an amorphous film should be detrimental to reliability of a high current device. Operating temperatures could easily reach the crystallization temperature of the film, thus drastically changing its electronic, thermal and mechanical properties. Perhaps the most significant side effect would be in the change in density between an amorphous and a crystalline film. The density of an amorphous material is typically 5% lower than its crystalline equivalent. This change in density could easily create voids in a layered or otherwise constrained film. It is possible that this is the cause of void formation observed by Peca, et al.[5] who fabricated Ni$_2$Si contacts with alternating layers of Ni and Si of the correct stoichiometry.

A more subtle difference among the various films is the result of the processing parameters during pulsed laser deposition. This difference can be quantified by examining the macroparticles that are present on the films. These macroparticles are in the form of droplets of the nickel silicon eutectic closest to Ni$_2$Si. This eutectic is 24.5 atomic percent silicon. **Figure 2** shows a SEM micrograph showing the dispersion of droplets on a smooth film from sample C. For the most part, the smaller droplets are spherical caps, while the larger droplets have a depressed center region giving them a doughnut-like appearance. There are visible differences between the droplets on films grown at room temperature and those on films grown with elevated substrate temperatures. For room temperature films, the droplets have a more pronounced center depression, while the droplets on films grown at high temperature appear more spherical as if the whole droplet was in the process of being absorbed by the film. This difference is shown in **Figure 3** and is quantified in **Table I** where the statistics between the droplets distributions of the various films are given.

Figure 2. SEM Micrograph of Ni_2Si surface on 4H-nSiC (Sample C) deposited with substrate temperature of 500°C.

The effect of laser fluence can be inferred by comparing sample A and sample B. Sample B has an areal density per unit thickness that is slightly smaller than that of sample A, but a percent coverage per unit thickness that is more than two and a half times larger than that for sample A. This difference is most likely caused by the high fluence ($10 \ Jcm^{-2}$) that vaporizes a larger portion of the ejected target material, thus producing a finer dispersion of the droplets.

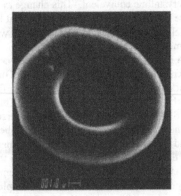

Figure 3 SEM micrographs comparing droplets of Ni_3Si on films of Ni_2Si deposited at 500°C (left) and room temperature (right)

Table I. Film structure and macroparticle distributions compared to sample preparation for silicon carbide substrates.

	Sample A	Sample B	Sample C
System	Compex 205	LPX 300	LPX 300
Substrate temp. (°C)	RT	RT	500
Fluence (Jcm^{-2})	10	2.3	2.3
Microstructure	Nanocrystalline-amorphous	Nanocrystalline-amorphous	Polycrystalline
Film thickness (nm)	150	1050	600
Droplet density /thickness (No.mm^{-2}nm^{-1})	9.06	7.83	0.146
Mean droplet diameter (μm)	2.32	3.86	2.62
Largest droplet diameter (μm)	14.46	14.68	9.0
Percent coverage / thickness (nm^{-1})	3.8×10^{-5}	9.2×10^{-5}	7.9×10^{-7}

The lower fluence deposition (sample B, 2 Jcm^{-2}) results in a larger portion of the melt that is not vaporized thus producing a smaller number of larger area droplets on the substrate. SEM micrographs of the target surface indicate significant melting for both fluence levels.

A comparison between sample C and sample B shows the effect of substrate heating. The effect of substrate heating on the deposited film microstructure has already been shown, however, there also seems to be an additional interaction between the droplets on the heated substrate that is not present in the films that are grown at room temperature. For each of the deposition conditions examined, the heated substrate had a reduced presence of droplets. It is as if the droplets were absorbed into the film after colliding with the substrate. This is apparent in the severely reduced droplet density, mean droplet diameter and diameter of the largest droplet.

There is a negligible difference between the electrical performance of the films grown at room temperature and 500°C as shown in **Figure 4**. In both cases the films exhibit "near ohmic" performance.

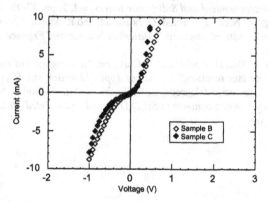

Figure 4 Comparison of the I-V electrical behavior between Sample B (open diamonds) and Sample C (filled diamonds).

That is, the I-V curve is non-rectifying, but there is not a linear relationship that one expects of a true resistive ohmic contact. However, the structural analysis of these films show that from a reliability perspective depositing Ni_2Si onto SiC at elevated temperatures is more desirable than depositing at room temperature owing to the more stable microstructure and the reduction of droplet size and droplet density.

CONCLUSIONS

This work has examined the structure effect of varying laser fluence and substrate temperature while depositing a specific nickel silicide onto single crystal silicon carbide. The particle density, mean particle diameter and maximum particle diameter were found to vary with both laser fluence and substrate temperature. From a device reliability perspective, the films grown at higher fluence and high substrate temperature were more desirable than films grown on unheated substrates owing to the crystalline nature of the resulting films and the reduction of particles on the film surface.

ACKNOWLDEGEMENTS

This work was supported through the joint Army Research Laboratory - Johns Hopkins University Advanced Materials Collaborative Research Program (DAAL01-96-2-0047).

REFERENCES
1.   M. Willander, H. L. Harnagel, and Ed., *High Temperature Electronics*. London: Chapman & Hall, 1997.
2.   M. W. Cole, C. Hubbard, C. G. Fountzoulas, D. Demaree, A. Natarajan, R. A. Miller, D. Zhu, and K. Xie, "The reliability of Ni contacts on n-SiC subjected to pulsed thermal fatigue," *Electrochemical and Solid-state letters*, vol. 2, pp. 97-99, 1999.
3.   A. Bachli, M.-A. Nicolet, L. Baud, C. Jaussaud, and R. Madar, "Nickel film on (001) SiC: Thermally induced reactions," *Materials Science and Engineering B*, vol. B56, pp. 11-23, 1998.
4.   I. Ohdomari, S. Sha, H. Aochi, and T. Chikyow, "Investigation of thin-film Ni/single crystal SiC interface reaction," *Journal of Applied Physics*, vol. 62, pp. 3747-3750, 1987.
5.   B. Peca, G. Radnoczi, S. Cassette, C. Brylinski, C. Arnodo, and O. Noblanc, "TEM study of Ni and Ni_2Si ohmic contacts to SiC," *Diamond and Related Materials*, vol. 6, pp. 1428-1431, 1997.

Mat. Res. Soc. Symp. Vol. 617 © 2000 Materials Research Society

PREPARATION OF SUPERHARD FUNCTIONALLY GRADED TETRAHEDRAL AMORPHOUS CARBON COATINGS BY PULSED LASER DEPOSITION

Q. WEI[*], S. YAMOLENKO[*], J. SANKAR[*], A. K. SHARMA[**], and J. NARAYAN[**]
[*]NSF Center for Advanced Materials and Smart Structures, Dept of Mechanical Engineering, McNair Hall, North Carolina A&T State University, Greensboro, NC 27411, quiming@ncat.edu
[**]Department of Materials Science and Engineering, Burlington Labs, P. O. Box 7916, North Carolina State University, Raleigh, NC 27695-7916

ABSTRACT

The internal compressive stress as large as 10 GPa has been the major stumbling block for preparation of relatively thick superhard tetrahedral amorphous carbon (Ta-C) films. We have successfully deposited Ta-C films as thick as 1000 nm by mechanical doping to reduce and alleviate the level of internal compressive stresses. In this paper, we reported the preparation of functionally graded Ta-C coatings by pulsed laser deposition. The thickness of the Ta-C films with significantly improved adhesion was measured to be up to 1500 nm. The concentration of foreign atoms such as silver, copper, silicon and titanium was decreased away from the coating/substrate interface, and the surface layer was pure Ta-C. Nanoindentation measurements were performed on the coatings. Nanohardness as high as 65 GPa and Young's modulus as large as 600 GPa were obtained for the functionally graded Ta-C films. Micro-Raman measurements and microstructural analysis by transmission electron microscopy was carried out to acquire information about the bonding environment and atomic structure of the coatings as a function of foreign atoms. The results were discussed in combination with theoretical models associated with the prediction of elastic properties of amorphous carbon networks.

INTRODUCTION

The amorphous state of carbon showed a wide spectrum of atomic structure and properties within the two extremes of crystalline diamond and graphite[1]. Most of the properties of amorphous carbon depend on the short-range environment of the atoms, especially on atomic coordination. For example, amorphous carbon that largely consists of tetrahedral bonding (sp^3) can have mechanical properties comparative to crystalline diamond (Young's modulus ~1000 GPa and hardness ~100 GPa), with a very low coefficient of friction, an optical band gap as large as 2.5 eV, considerable field emission effect, IR transparent, chemically inert, and so on. This type of amorphous carbon has been designated as diamondlike carbon (DLC), or tetrahedral amorphous carbon (ta-C). One more advantage associated with DLC is that its properties can be easily tailored by the ratio of the tetrahedral bonding population to the trigonal bonding population (sp^3/sp^2 ratio). Due to its unique physical, chemical and mechanical properties, DLC has found applications as antireflective coatings for IR windows[2], as electron emitters[3], as protective coatings for magnetic disk drives[4], etc. However, the advantages of DLC have been diminished by the large internal compressive stresses built up during preparation by whatever technique. The level of this internal compressive stress can be as high as 10 GPa[5] and can raise serious adhesion concern. Decohesion of thin coatings can occur when the internal stress, σ, exceeds a critical value which sets the upper limit of the thickness of the coatings. Traditional approaches to obtaining DLC films with relatively low internal stress levels involve increasing deposition temperatures or decreasing the energies of carbon species arriving at the substrate surface, *etc.* Unfortunately, all these are achieved at the expense of reducing sp^3/sp^2 ratio, which offsets against the advantages of DLC. Recently, Friedmann *et al.*[6] reported preparation of thick stress-free amorphous tetrahedral carbon films with hardness near that of diamond by pulsed laser deposition. They first deposited thin (100-200 nm) films at room temperature, and then annealed the films at 600°C for a short time (2 min) to relieve the large (6-8 GPa) internal compressive stresses. The films were then cooled down to room temperature for further deposition. They have produced adherent films up to 1.2 μm thick with low residual compressive stress (<0.2 GPa). Raman and electron energy loss spectra from single-layer annealed specimens show only subtle changes from as-grown films. Ferrari *et al.* [7] reported a comprehensive study of the stress release and structural changes caused by post-deposition thermal annealing of Ta-C on Si produced by filtered cathodic vacuum arc (FCVA) deposition, and reached similar conclusions. The advantage of thermal annealing is that pure DLC films can be prepared. However, one of the disadvantages is that the throughput is largely decreased, and for microelectronic applications the thermal budget may become a great concern.

We have reported successful preparation of relatively thick adherent DLC films by pulsed laser deposition through incorporation of relatively compliant elements such as metals [8]. In this paper, we report preparation and characterization of even thicker DLC coatings by functional grading (FG) design.

EXPERIMENT

We have adopted an ingenious modification of the traditional PLD process in order to incorporate foreign atoms into the growing films. Briefly, several targets were loaded, with one being pure graphite, and others being covered in part by a piece of the desired alloy element target. The target carousel of the deposition chamber can hold four targets at the same time. The alloy element was incorporated during PLD process, with the concentration of the alloy element decreased away from the coating/substrate interface. For the deposition of the layers between the top, pure DLC layer and the substrate, since the target was spinning, the focused laser beam would impinge sequentially on graphite and the alloy element piece to ablate the target materials to form composite layers. The Si (100) wafers were used as substrates, which were cleaned in acetone and methanol ultrasonic baths followed by HF dip to remove the native oxide layer before loading into the laser deposition chamber. Silver, copper and titanium were chosen as alloy elements to be incorporated into the composite layer.

The laser beam source used was KrF pulsed excimer laser (λ=248 nm, duration t_s=25 nm) at a repetition rate of 10 Hz, with an energy density close to 3.0 J/cm^2. All the depositions were conducted at room temperature in a high vacuum exceeding 1×10^{-7} torr. Three layers of different alloy element contents were deposited for 30 mins, and the final top pure DLC layer was deposited for 30 mins. The thickness of the coatings was measured with an optical profilometer (Wyko Corporation). The coatings were analyzed using micro-Raman spectroscopy for bonding characteristics. Transmission electron microscopy (TEM) (cross-sectional) was used to study the microstructures of the coatings. The TEM was TOPCON002B operated at 200 KV at a point-to-point resolution of 1.8 Å at the first Scherzer focus. Nano-mechanical properties of the coatings were characterized using Nano-indenter XP^{TM}.

RESULTS AND DISCUSSIOIN

It was found by optical profilometer that the thickness of the FG diamondlike carbon coatings is from 1.4 to 1.6 μm, depending on the type of alloy element incorporated below the pure DLC layer. This is understood on the basis of the different ablation rate of the selected alloy elements. We did not measure the absolute value of the alloy element contents, but from previous work of doped DLC films the estimate of the maximum alloy element content in the coating does not exceed 4.0 at% for Ag and Cu, and does not exceed 7.0 at% for Ti.

Fig. 1 shows the visible micro-Raman spectra of the doped part of the FG DLC coatings. The visible Raman spectra exhibit typical Raman features for amorphous carbon, with no evidence of significant change of short-range order of the atomic structure upon incorporation of these foreign atoms. All the spectra show a broad hump centered at around ~1560 cm^{-1} which extends to ~1350 cm^{-1}. The first part of the Raman spectra located at ~1560 cm^{-1} corresponds to the G-peak of crystalline graphite, and the second part (a shoulder) at ~1350 cm^{-1} corresponds to the D-peak of very small graphite crystallites. The insignificant shoulders in the Raman spectra indicate that there are few small graphite crystallites in the films.

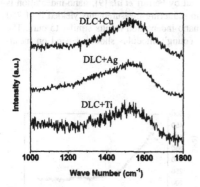

Fig. 1 Visible micro-Raman spectra of DLC layers that contain different types of alloy elements.

In order to understand the microstructure of the FG DLC coatings, cross-sectional TEM was performed on some samples. One of the problems associated with cross-sectional TEM specimen preparation of superhard DLC coatings is that the ion-milling rate of the superhard coating is extremely slow as compared with the Si substrate. Therefore, there is always a bridge of the coating material that is very thick while the substrate material has been milled away. This makes it difficult to study the interface between the coating and the substrate, except that special technique is utilized to prepare the TEM specimen. Fig. 2 (a) shows the bright field TEM image of the cross section of the FG DLC coating (two coatings glued together face-to-face). The top layer of the coating is pure DLC, while the layers in between are DLC with different contents of copper (concentration of copper decreases away from the interface). Selected area diffraction patterns exhibit no evidence of crystalline states (Fig. 2 (b)). Fig. 2 (c) is the high resolution TEM image of the coating, showing discontinuous layered structure and small clusters presumably of copper.

We have reported the effect of foreign element incorporation on internal compressive stress reduction of DLC coatings prepared by PLD and have observed significant decrease of stress

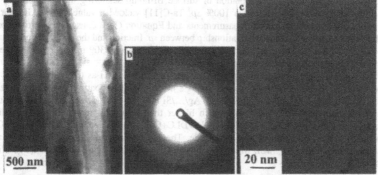

Fig. 2 Bright field TEM image of the cross-section of the DLC/DLC+Cu/.../Si FG coating (a), selected area diffraction pattern showing no crystalline phase (b), and high resolution TEM imgae showing discontinuous layered structure (c) and small clusters presumably of copper.

level due to the presence of foreign atoms such as Cu, Ti, and even Si[8]. Since in the case of Si incorporation, we observed considerable population of Si particulates in the films, and we also found that the incorporation of such alloy elements reduced the nano-hardness and elastic modulus, in this research, we replaced Si by Ag, and employed the functional grading design.

In order to understand the mechanical behavior of the FG coatings, we performed nano-indentation measurements on the coatings. Fig. 3 is the nano-hardness (a) and Young's modulus (b) as a function of indentation depth for a single layer of pure DLC deposited with 309-mJ laser energy. Values of ~420 GPa for Young's modulus and ~54 GPa for nano-hardness are obtained for this pure thin DLC film. However, as pointed out by Ferrari et al. [9], nano-indentation is seen to underestimate E for Ta-C. Surface Brillouin scattering measurements yielded E of 750 GPa for FCVA deposited Ta-C films which give nano-indentation results similar to ours. The constraint-counting network[10] predicts that the Young's modulus should depend on mean-atomic coordination Z as

$$E=E_0(Z-Z_0)^{1.5}, \qquad (1)$$

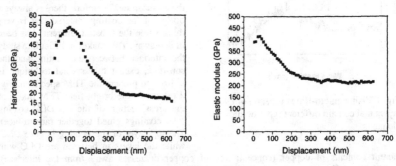

Fig. 3 Nano-hardness (a) and Young's modulus (b) of pure DLC layer produced by 309 mJ laser energy. The film thickness is about 300 nm. The substrate effect on the nanoindentation is clear.

where Z_0 is the critical coordination, below which the networks have zero rigidity. Theoretical value of Z_0 was given to be 2.4, and Ferrari et al.[9] obtained a value of 2.6 for Z_0 based on experimental observations. Extrapolation of surface Brillouin scattering measurements[9] and molecular dynamics simulation of a 100% sp^3 Ta-C[11] yielded a value of 480 GPa for E_0. Therefore, from nano-mechanical measurements and Equation (1), a conservative estimation of sp^3 fraction can be made using the relationship between sp^3 fraction and the coordination number, which is $Z = 3 + sp^3$. Therefore, for the pure DLC deposited with 309 mJ laser energy, the sp^3 fraction is conservatively estimated to be ~51%.

Fig. 4 (a) is the nano-hardness of FG DLC with Ag, Cu and Ti as the alloy elements. The curves showed the same tendency as the Young's moduli of these coatings, which are given in Fig. 4 (b). The nano-hardness of DLC/DLC+Ag/.../Si is about 60 GPa, which is close to that of crystalline diamond (~100 GPa), and much higher than superhard materials such as silicon carbide (~35 GPa). The nano-hardness of DLC/DLC+Cu/.../Si FG coating is close to ~50 GPa, also much higher than that of silicon carbide. The FG coating that shows the lowest nano-hardness is DLC/DLC+Ti/.../Si, which is only ~40 GPa, still slightly higher than that of silicon carbide.

Fig. 4 (b) gives the Young's moduli for FG DLC with Ag, Cu and Ti as the alloy elements. It is seen that the FG coating of DLC/DLC+Ag/.../Si exhibits the highest value of Young's modulus. From the elastic modulus and Equation (1), we can estimate the sp^3 fraction of the pure DLC layer to be ~64%. For DLC/DLC+Cu/.../Si FG coating, the estimated sp^3 fraction for the top pure DLC layer is about 53%, and for DLC/DLC+Ti/.../Si FG coatings, it is about 45%. Our previous work showed that incorporation of Ti into the DLC coatings reduces the Young's modulus more than Cu does[8]. It was also found that Ti form metallic Ti-C bonding in the film[12]. Neither copper nor silver form strong chemical bonds with carbon, and the existence of these elements may only help the relaxation of the internal compressive stress.

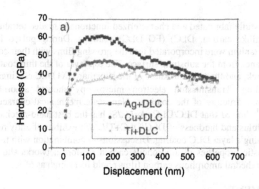

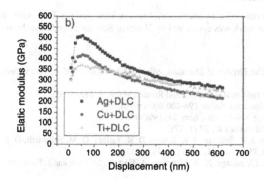

Fig. 4 Nano-hardness (a) and elastic modulus (b) as a function of indentation depth (displacement) of functionally graded DLC coatings with Ag, Cu and Ti as the alloy elements. Notice the difference in the effect of the three different alloy elements (Ag, Cu and Ti) and decrease of substrate effect as compared to pure, thin DLC film case given in Fig. 3.

One of the advantages of functionally graded DLC coatings is the decrease of the substrate effect as reflected from the nano-indentation measurements, as one can observe by comparing

Fig. 3 against Fig. 4. In the case of pure DLC coatings, as shown in Fig. 3, the substrate effect sets in at the very early phase of nano-indentation. While in the case of thick FG coatings, the situation is significantly improved. Of course, the major advantage of the FG DLC design is that it allows us to produce much thicker DLC coatings than traditional approaches.

Voevodin et al. [13,14], also produced FG DLC coatings, but with a different route. They used a hybrid of magnetron sputtering and PLD multi-source scheme to deposit crystalline Ti, TiC and amorphous DLC films, with a total thickness of 2-3 µm and hardness around ~65 GPa. It is reasonable to expect that FG DLC design will be advantageous at least for tribological applications.

CONCLUSIONS

We have successfully fabricated and characterized functionally graded tetrahedral amorphous carbon (or diamondlike carbon, DLC) (FG DLC) coatings. During pulsed laser deposition, copper, silver and titanium were incorporated into the growing films with their concentration as a function of the distance from the substrate surface. The top layer of the thin coating is pure DLC of about 400 nm thick, and the total thickness of the superhard FG DLC coatings can exceed 1.0 µm without buckling. Transmission electron microscopy showed amorphous state and discontinuous layered structure of the coatings. Nanoscale mechanical characterizations using Nanoindenter XP^{TM} showed that DLC/DLC+Ag/.../Si has the best nano-mechanical properties, and the elastic modulus and hardness of DLC/DLC+Ti/.../Si exhibits slightly reduced values as compared to pure, single layer DLC coating. Discussions in combination with theoretical models that predict the elastic properties of amorphous carbon random networks showed that the sp^3 fraction in these tetrahedral amorphous carbon films could be as high as 65%.

ACKNOWLEDGMENTS

The authors would like to thank Dr. Minseo Park for his help with the micro-Raman measurements. This work was sponsored by National Science Foundation through NSF-CAMSS.

REFERENCES

1. Y. Lifshitz, in The Physics of Diamond, edited by A. Paoletti and A. Tucciarone, Italian Phy. Soc., 209 (1997).
2. A. Bozhko, A. Ivanov et al. Diamond and Related Mater. 4, 488 (1995).
3. W. I. Milne, J. Non-cryst. Solids 198-200,605 (1996).
4. J. Robertson, Prog. Solid State Chem. 21, 199(1991).
5. D. Nir, Thin Solid Films 146, 27 (1987).
6. T. A. Friedmann, J. P. Sullivan, J. A. Knapp, D. R. Tallant, D. M. Follstaedt, D. L. Medlin and P. B. Mirkarimi, Appl. Phys. Lett.71, 3820 (1997).
7. A. C. Ferrari, B. Kleinsorge, N. A. Morrison, A. Hart, V. Stolojan and J. Robertson, J. Appl. Phys.85, 7191 (1999).
8. Q. Wei, R. J. Narayan, A. K. Sharma, J. Sankar and J. Narayan, J. Vac. Sci. Tech. A17, 3406 (1999).
9. A. C. Ferrari, J. Robertson, M. G. Beghi, C. E. Bottani, R. Ferulano and R. Pastorelli, Appl. Phys. Lett. 75, 1893 (1999).
10. H. He and M. E. Thorpe, Phys. Rev. Lett. 54, 2107 (1985).
11. P. C. Kelires, Phys. Rev. Lett. 73, 2460 (1994).
12. Q. Wei, R. J. Narayan, J. Narayan, J. Sankar and A. K. Sharma, Mat. Sci. Eng. B53, 262 (1998).
13. A. A. Voevodin, M. A. Capano, S. J. P. Laube, M. S. Donley and J. S. Zabinski, Thin Solid Films 298, 107 (1997).
14. A. A. Voevodin and J. S. Zabinski, Diamond and Related Mat. 7, 463 (1998).

Mat. Res. Soc. Symp. Vol. 617 © 2000 Materials Research Society

Preparation of Crystalline Chromium Carbide Thin Films Synthesized by Pulsed Nd:YAG Laser Deposition

Kazuya DOI, Satoshi HIRAISHI, Hiroharu KAWASAKI, Yoshiaki SUDA,

Department of Electrical Engineering, Sasebo National College of Technology, Okishin 1-1, Sasebo, Nagasaki 857-1193 , Japan

ABSTRACT

Chromium carbide thin films are synthesized on Si(100) substrates by a pulsed Nd:YAG laser deposition (PLD) method as parameters of methane gas pressure. Glancing-angle X-ray diffraction patterns show that the film prepared by PLD method is a polycrystalline thin film composed of Cr_3C_2 and Cr_7C_3, even in the base pressure. Diffraction patterns, however, are depended on the methane gas pressure. Grain size of the prepared film increases with increasing methane gas pressure. One of the reasons of these phenomena may be considered to the phase reaction between the ablated species, such as Cr, CrCx and CH_4 gas in the plasma plume.

INTRODUCTION

Crystalline chromium carbide (Cr_3C_2) is one of the most popular carbides used to fabricate wear resistant surface films and in steel industries, because of its excellent physical properties, such as high melting temperature, strength, hardness and corrosion resistance.[1~2] However, there have not been many studies regarding this topic. Donley et al. prepared CrC_x thin films using a KrF excimer laser deposition method at the Si substrate temperature of room temperature and 300°C. However, both of the films were amorphous,[3] and high quality crystalline Cr_3C_2 films were not prepared. We also prepared the Cr_3C_2 films using a pulsed Nd:YAG laser deposition (PLD) method[4-9], which is a versatile method for deposition of crystalline thin films, such as superconducting, semiconducting, and ferroelectric films. Our previous results suggest that the substrate temperature is one of the most important parameters in the fabrication of crystalline chromium carbide film, and the film prepared at Ts ≥ 500 °C is a polycrystalline thin film composed of Cr_3C_2 and Cr_7C_3.[10] However, the growth mechanism of the Cr_7C_3 component was not well understood.

In this paper, we described the fundamental characteristics of the chromium carbide films prepared using the PLD method, and the effects of methane gas pressure on the properties of CrC_x thin films. On the basis of them, we discuss about the mechanism of the preparing crystalline Cr_3C_2 thin film.

EXPERIMENTAL

The schematic of the experimental apparatus is shown in Fig. 1. A deposition chamber was made of stainless steel with a diameter of 400 mm and a length of 370 mm. The chamber was evacuated to a base pressure (below 4.0×10^{-4} Pa) using a turbo molecular pump and a rotary pump. The gas pressure was varied from the base pressure to 10 Pa by feeding pure methane (CH_4) gas into the chamber. A pulsed Nd:YAG laser (Lumonics YM600; wavelength of 532 nm, pulse duration of 6.5 ns, maximum output energy of 340 mJ) was used to irradiate Cr_3C_2 (purity 99.9%) targets. Their radiated area was kept at 2.8 mm^2. The laser energy density (Ed) was fixed at 3.8 J/cm^2. The targets were rotated at 20 rpm to avoid pitting during deposition. The Si (100) substrates were cleaned using an ultrasonic agitator in repeated baths of ethanol and then rinsed in high-purity deionized water prior to loading into the deposition chamber. The substrates were located at a distance of 60 mm from the facing target and were heated to 700°C using an IR lamp. After 36000 laser pulses at a 10 Hz repetition rate, the deposition process was completed. The film thickness, measured by α-step (KLA Tencor; AS500), was about 210 nm and the growth rate was 0.12 nm/s. Details of the preparation conditions of chromium carbide thin films are given in Table I.

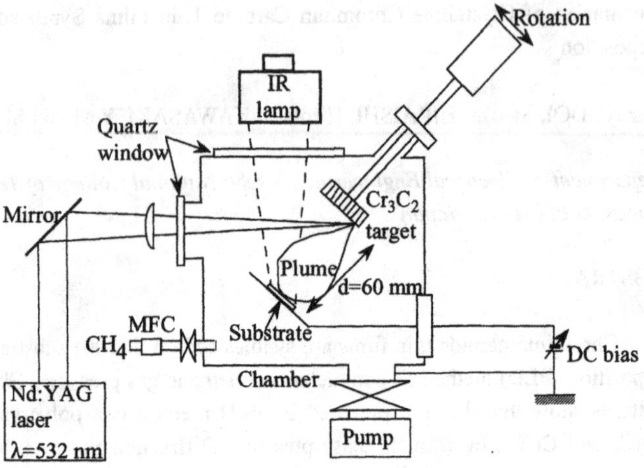

The surface morphology was observed using a field-emission secondary electron microscope (FE-SEM; JEOL JSM-6300F). The compositions of the CrC_x films were examined using an X-ray Photoelectron Spectroscopy (XPS; Shimadzu ESCA-850M). The crystalline structure and crystallographic orientation of the films were characterized by glancing-angle X-ray diffraction (GXRD; PHILIPS PW1350) using Cu Kα radiation where the angle of incidence was kept at 1.0°.

Fig. 1   Experimental setup.

Table I   Preparation conditions

Laser	Pulsed Nd:YAG laser Wavelength 532 nm Pulsed width 6.5 ns Energy density 3.8 J/cm^2 Repetition rate 10 Hz
Target	Cr_3C_2 (purity 99.9%)
Rotating speed	20 rpm
Substrate	Si (100)
Substrate temp.	room temp ~ 700°C
Distance	d = 60 mm
Base pressure	< 4.0×10^{-4} Pa
Gas pressure	4.0×10^{-4}~10 Pa (CH_4)
Deposition time	30 min

RESULTS AND DISCUSSION

The surface morphology of the CrC_x thin films on Si (100) substrates was examined using FE-SEM at various substrate temperatures (Ts). Figure 2(a) and 2(b) show the micrographs of the film prepared at substrate temperature (Ts) of 700 °C. Fig. 2(a) shows the micrograph of its magnification of 5000. The film is smooth and pinhole-free, but there are several droplets. Fig. 2(b) shows the micrograph of its magnification of 100000. The film is consists of many grains of which sizes are about 20~80 nm.

(a)   Magnification of 5000　　　　　　(b)   Magnification of 100000

Fig. 2   FE-SEM micrographs of the CrCx film prepared at Ts = 700 °C
(P_{CH4}=10Pa).

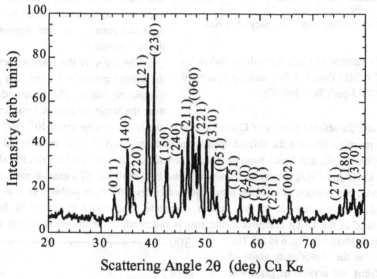

Fig. 3   XRD pattern of a reference target pellet of Cr_3C_2.

Detailed GXRD measurements were conducted to study the crystalline properties of the
laser-deposited chromium carbide films. Figure 3 shows the XRD pattern of a reference target pellet
of Cr_3C_2, in which numerous crystalline peaks of Cr_3C_2 were identified. Figure 4(a)~4(c) show the
XRD patterns of films deposited as a parameter of methane gas pressure. All of the films are
polycrystalline composed of Cr_3C_2 and Cr_7C_3 independent of methane gas pressure. There are
several peaks of crystalline Cr_3C_2 and Cr_7C_3, even in the base pressure as shown in Fig. 4(a), unlike
Fig. 3. With increasing the methane gas pressure (P_{CH4}), however, XRD pattern of the film is
changed. One of the reasons of this phenomenon may be considered as follows. There are two

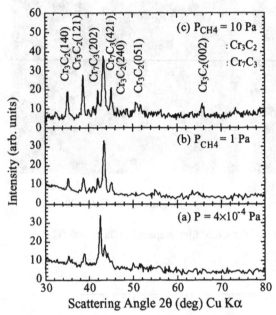

Fig. 4 XRD patterns of CrC_x films deposited at (a) $P=4\times10^{-4}$ Pa, (b) $P_{CH4} = 1$ Pa and (c) $P_{CH4} = 10$ Pa ($E_d = 3.8$ J/cm^2, $T_s = 700$ °C).

mechanisms of creation Cr_3C_2 and Cr_7C_3 components. One of them is the reaction on the substrate surface. In our previous study, the ablated species from the Cr_3C_2 target by Nd:YAG laser are estimated to be mainly Cr, C atoms and ions, and related molecules such as Cr_3C_2 and Cr_7C_3 and so on.[11] In general, the boiling energy of the carbon (4830 °C) is much higher than that of chromium (2482 °C). Therefore, the density of Cr atoms may be higher than that of C atoms in the plasma plume, and thus, the density of Cr atoms related to the surface reaction is higher than that of C atoms on the surface of the substrate. As this reaction may be dominant in the base pressure, Cr_3C_2 and Cr_7C_3 components considered to be appeared even in the film deposited at the base pressure.

The other is the reaction between the ablated species of Cr and CH_4 gas in the plasma plume. The plasma density near the target surface is considered to be very high, and the ablated species of C and CrC_x near the target may be about $10^{14}\sim10^{15}$ cm^{-3}. Therefore, the reactions between the ablated species and methane gas (CH_4) may be occurred in the plasma plume, and Cr_3C_2 and Cr_7C_3 components are generated. The reaction between the ablated species of Cr, CrC_x and CH_4 gas may be dominant in the $P_{CH4} = 10$ Pa. XPS measurement shows that the C/(Cr+C) composition ratio of the film deposited at $P_{CH4}=10$ Pa is higher than that of the film deposited at base pressure. In this system, substrate temperature is important for the crystallinity grade of the Cr_3C_2 component, but that is independent for the XRD pattern of the deposited films as shown in Fig. 6 in ref. 10.

To determine the crystallinity grade of the prepared films, the crystalline grain size was determined from the Full Width Half Maximum (FWHM) of the X-ray peaks of $Cr_3C_2(121)$, as shown in Fig. 4(a)~(c), throughout the equation:

$$\text{grain size} = \frac{0.9 \cdot \lambda}{\cos\theta \cdot \text{FWHM}}$$

where λ is the wavelength of the incident radiation and θ is the Bragg's angle. The obtained values are shown in Fig. 5. The

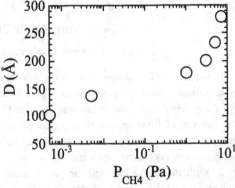

Fig. 5 Crystal grain size of $Cr_3C_2(121)$ estimated from Fig. 4.

grain size of $Cr_3C_2(121)$ is increased with increasing P_{CH4}. These results suggest that the CH_4 gas pressure is one of the important parameters in the fabrication of the crystalline Cr_3C_2 thin films.

CONCLUSION

Chromium carbide thin films are prepared on Si(100) substrates by PLD method as parameters of methane gas pressure. The deposited films are smooth and pinhole-free. Glancing-angle X-ray diffraction shows that the prepared films are polycrystalline composed of Cr_3C_2 and Cr_7C_3, even in the base pressure. However, diffraction patterns are strongly depended on the P_{CH4}. Grain size of the Cr_3C_2 component of prepared film increases with increasing P_{CH4}. One of the reasons of these phenomena may be due to the phase reaction between the ablated species, such as Cr, CrCx and CH_4 in the plasma plume.

ACKNOWLEDGEMENTS

This work was supported in part by Grant-in-Aid for Scientific Research (B), the Regional Science Promoter Program, the Shimadzu Science Foundation and a Research Fund from the Nagasaki Super Technology Development Association. The authors wish to thank Drs. K. Ebihara, T. Ikegami and Y. Yamagata of Kumamoto University and Dr. A. M. Grishin of the Royal Institute of Technology for their helpful discussions. The authors also wish to thank Dr. H. Abe and Mr. H. Yoshida of the Ceramic Research Center of Nagasaki for their technical assistance with the experimental data.

REFERENCES

1)   E. Klar: *Metal Handbook*, (American Society for Metals, Metal Park, Ohio, June, 1984) 9th ed., Vol. 7, p. 804.
2)   K. Isozaki, Y. Hirayama and Y. Imamura: US Patent **492** (1990) 7791.
3)   M. S. Donley, J. S. Zabinski, W. J. Sessler, V. J. Dyhouse, S. D. Walck and N. T. McDevitt: Mater. Res. Soc. Symp. Proc. **236** (1992) 461.
4)   Y. Suda, H. Kawasaki, R. Terajima and M. Emura: Jpn. J. Appl. Phys. **38** (1999) 3619.
5)   Y. Suda, T. Nakazono, K. Ebihara and K. Baba: Thin Solid Films **281-282**   (1996) 324.
6)   Y. Suda, T. Nakazono, K. Ebihara and K. Baba: Nucl. Instrum. & Methods in Phys. Res. **B121** (1997) 396.
7)   Y. Suda, T. Nakazono, K. Ebihara, K. Baba and H. Hatada: Mater. Chem. & Phys. **54** (1998) 177.
8)   Y. Suda, H. Kawasaki, R. Terajima, M. Emura, K. Baba, H. Abe, H. Yoshida, K. Ebihara and S. Aoqui: J. Korea. Phys. Soc. **35** (1999) S88.
9)   Y. Suda, T. Nakazono, K. Ebihara, K. Baba and S. Aoqui: Carbon **36** (1998) 771.
10)   Y. Suda, H. Kawasaki, R. Terajima and M. Emura: Jpn. J. Appl. Phys. **38** (1999) 3619.
11)   K. Murakami, H. Asako, T. Okamoto and Y. Miyamoto: Mater. Sci. Eng. **A123** (1990) 261.

AUTHOR INDEX

SUBJECT INDEX

Printed in the United States
By Bookmasters